NURB Curves and Surfaces

For Dianne

NURB Curves and Surfaces

From Projective Geometry to Practical Use

Gerald E. Farin
Arizona State University
Tempe, Arizona

A K Peters
Wellesley, Massachusetts

Editorial, Sales, and Customer Service Office

A K Peters, Ltd.
289 Linden Street
Wellesley, MA 02181

Library of Congress Cataloging-in-Publication Data

Farin, Gerald E.
NURB curves and surfaces : from projective geometry to practical use / Gerald Farin.
p. cm.
Includes bibliographical references.
1-56881-038-5
1. Spline theory —Data processing. 2. Computer-aided design.
I. Title.
QA224.F37 1994 94-25631
511'.42—dc20 CIP

Cover illustration from *The Complete Woodcuts of Albrecht Dürer*, edited by Dr. Willi Durth, Dover Publications, Inc., New York, 1963. Plate 338 taken from *Illustrations to the Treatise on Measurement*, Unterweysung der Messung, Nürnberg, 1525. A man drawing a lute.

Printed in the United States of America
98 97 96 95 94 10 9 8 7 6 5 4 3 2 1

Contents

Preface

Computers have become an integral part of today's design and manufacturing technology. Consequently, this field now employs sophisticated mathematical methods: mathematics is the only language "understood" by computers. One aspect of the design process — modeling complex parts such as car bodies or airplane fuselages — has given rise to techniques that are now known by the names Bézier and Coons methods. The corresponding theories, together with that of B-splines, constitutes what is now referred to as "curve and surface design."

Until quite recently, different companies used different variations of these basic techniques. To name just three: General Motors used Gordon surfaces, Ford used Coons patches, and Chrysler used Chebychev polynomials. This diversity of methods led to communication problems: a product definition in one form is typically not translatable into another system. Thus a *geometry standard* was called for, and it was found in the form of NURBS (short for NonUniform Rational B-Splines, a term that will be explained later).

I find it very instructive to look at the development at the Boeing Co.: during the fifties, engineers "computerized" existing design and drafting methods which were all based on the use of conics.[1] In the late fifties,

[1] For an excellent exposition of these methods, see the book by Liming [78]. This book is the first to introduce algorithms into the design process — although in those days, CAD would probably have been translated as "Calculator Aided Design."

J. Ferguson developed a spline package for Boeing's design software. Both packages were independently used, and eventually it was discovered that they were essentially incompatible. That is to say, a spline is not a conic, and a conic is not a spline, with the somewhat trivial exception of the parabola. Management realized that these different methods had to be unified, and thus NURBS were born: they encompass both splines and conics.

The first appearance of NURBS is in K. Vesprille's PhD thesis [115], where he makes heavy use of homogeneous coordinates, an approach that goes back to S. Coons and R. Forrest. Riesenfeld [99] realized early that using homogeneous coordinates mandates working with projective geometry. This book follows these ideas and attempts to base the theory of NURBS firmly in projective geometry. This way, we gain more insight into theoretical issues as well as practical ones.

After a general outline of projective geometry, we will introduce conics through a classical projective definition. Using the concept of cross ratios, we arrive at the de Casteljau algorithm for conics. Then we cover areas such as rational Bézier curves, NURB curves and surfaces, triangular patches, Gregory patches, and more. The book closes with some practical examples, including a discussion of the IGES NURBS data specifications.

I have taught the material in this book in one semester classes, typically at the beginning graduate level. The prerequisites are linear algebra, calculus, and some basic computer graphics. The latter is needed for successful completion of programming assignments, which I find indispensable for a true comprehension of the subject. Such exercises, both practical and theoretical, are included at the end of each chapter.

This book would not have been possible without the help of many friends and colleagues. I would like to thank R. Barnhill, K. Choodamani, L. Gross, D. Jung, P. Kashyap, T. Kim, A. Razdan, A. Palacios, T. Sederberg, R. Smith, A. Swimmer, H. Wolters, and A. Worsey for their many contributions. The most support, however, came from D. Hansford, who carefully monitored the progress of this work. I also acknowledge support from the National Science Foundation through Grant No. DM9123527 and a grant of the German Research Council (DFG).

Gerald Farin
Lingen/Ems

1

The Projective Plane

1.1 Motivation

Most of us have studied geometry to some extent once, and in most cases, it was *Euclidean geometry.* That geometry is perfectly adequate to handle objects such as points and lines, circles and planes, and much more. In fact, it is used successfully in the computer aided construction of cars, ships, and airplanes – the process known as CAD, or Computer Aided Design.

As we shall see later, projective geometry is also the natural setting for NURBS: many of their intrinsic properties are much more easily understood in a projective context.

Euclidean geometry is an indispensable tool for some applications, but it has its shortcomings in other areas. If we are to deal with how we *see* things, the Euclidean viewpoint becomes less advantageous. Looking up at highrise buildings, we perceive parallel lines as converging, and Euclidean geometry can't cope with that. In projective geometry, there are no parallel lines — it is therefore the geometry of choice for those working with perceived images. Architects use projective geometry when drawing a building as it would appear to an observer; computer graphics programmers use it when they model "realistic" scenes.

The roots of projective geometry go back to the late Middle Ages; and they were not in mathematics, but in the arts. Medieval painters neglected the concept of realistic perspective for religious reasons. It was the advent of the Renaissance that led artists to study the techniques necessary for realistic rendering. Among them, we find names such as P. Uccello,[1] Leonardo da Vinci, and Albrecht Dürer. The first mathematical excursions into the field of perspective were carried out by B. Pascal and G. Desargues in the seventeenth century. From then on, perspective was an indispensable tool for artists, architects, and engineers; but it took until the nineteenth century that it became the object of formal mathematical inquiry, culminating in the theory of projective geometry. Some of the names associated with this breakthrough are J. Poncelet, K. von Staudt, J. Steiner, and F. Moebius.

Figure 1.1. Perception of an object: the observer "sees" the object's projection into a plane. Woodcut by A. Dürer, 1525.

Before giving formal definitions in the next section, let us discuss the motivation behind the underlying principle of projective geometry. Remember, this geometry was invented to model how we perceive objects. So let's assume that an observer is looking at a three-dimensional object.

[1]The cover of the journal "Computer Aided Geometric Design" is one of Uccello's studies of perspective renditions.

The observer's perceived image of the object is two-dimensional: it is the projection of the object into a plane, as shown in Figure 1.1.

Now for a more formal treatment: place the observer at the origin of a 3-D orthogonal coordinate system, and let $z = 1$ denote the plane into which we project. Let us make our idealized observer look at some 3-D point $\mathbf{x}$. The perceived image is $\mathbf{x}$'s projection into the plane $z = 1$. But many more 3-D points project onto the same image: all these points, located on the straight line through the origin and $\mathbf{x}$, are indistinguishable to our observer! Figure 1.2 illustrates. In projective geometry, since we can't tell those points apart anyway, they are treated as *one point.*

This idea — only treating as an entity what you can identify after you project it into the plane — is carried over to all geometrical objects. Let's take the case of straight lines, or lines for short. Referring to Figure 1.3, we see that all lines contained in the indicated plane project onto the same line in the plane $z = 1$. Consequently, they are treated as being identical! This plane is characterized by its normal vector, and so that normal vector is used to denote the equivalent lines.

Our treatment of projective geometry will not be exhaustive – we just need it as a tool for the description of rational curves and surfaces. For more details, the interested reader should consult the texts by Boehm/Prautzsch [25], Coxeter [33], Hilbert/Cohn-Vossen [66], Pedoe [87], Penna/Patterson [88], or Struik [111].

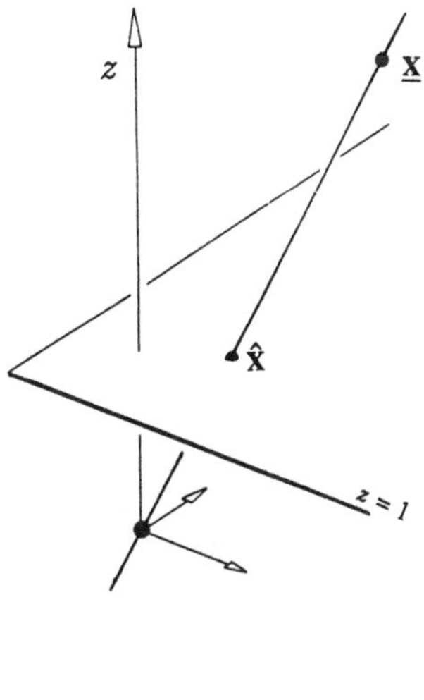

Figure 1.2. Perception of a point: all points on the shown line appear identical to an observer at the origin.

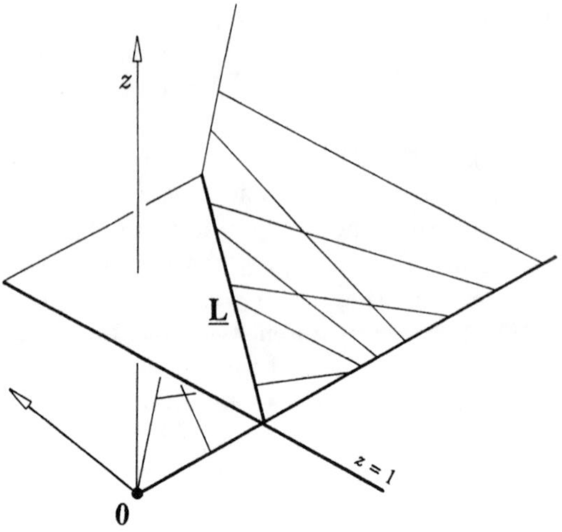

Figure 1.3. Perception of a line: all lines in the indicated plane are identical to an observer at the origin.

1.2 The Projective Plane

The intuitive concepts from the preceding section lead to the following formal definition of the projective plane $\mathbb{P}^2$:

> The **projective plane** $\mathbb{P}^2$ consists of points $\underline{\mathbf{p}}$, denoted by coordinate columns
>
> $$\underline{\mathbf{p}} = \begin{bmatrix} p_1 \\ p_2 \\ p_3 \end{bmatrix}$$
>
> and lines $\underline{\mathbf{L}}$, denoted by coordinate rows
>
> $$\underline{\mathbf{L}} = [l_1, l_2, l_3].$$
>
> All nonzero multiples $\alpha\underline{\mathbf{p}}$ denote the same location, and we define $\underline{\mathbf{p}} \hat{=} \alpha\underline{\mathbf{p}}$. Similarly, all nonzero multiples $\alpha\underline{\mathbf{L}}$ are treated as the same line, and we set $\underline{\mathbf{L}} \hat{=} \alpha\underline{\mathbf{L}}$.
> A point $\underline{\mathbf{p}}$ is on a line $\underline{\mathbf{L}}$ if
>
> $$\underline{\mathbf{L}}\underline{\mathbf{p}} = 0, \tag{1.1}$$
>
> i.e., if the dot product of their coordinates vanishes.

These definitions invite several comments:

1. The coordinate column $[0,0,0]^{\mathrm{T}}$ designates a point, but not a *location*: as it lies on every line, its location is indeterminate. Similarly, the coordinate row $[0,0,0]$ designates a line, but its location is indeterminate as every point lies on it.

2. We had used the plane $z = 1$ to introduce $\mathbb{P}^2$, but now points with $z = 1$ are not special. We will treat them as special later when we talk about affine geometry.

3. The projective plane does not possess "natural" coordinate axes; in particular, it does not possess an origin. (In particular, $[0,0,0]^{\mathrm{T}}$ is not a candidate for an origin, since it does not even designate a location.)

4. The sign "=" denotes algebraic equivalence, while "$\hat{=}$" denotes geometric equivalence.

5. We may replace point coordinates by any nonzero multiple without changing the point's location: $\underline{\mathbf{x}} \hat{=} \underline{\mathbf{y}}$ implies $\underline{\mathbf{x}} = \alpha\underline{\mathbf{y}}$. But we may *not* change point coordinates within expressions: $\underline{\mathbf{x}} = \underline{\mathbf{y}} + \underline{\mathbf{z}}$ does not imply $\underline{\mathbf{x}} = \underline{\mathbf{y}} + \alpha\underline{\mathbf{z}}$. This is also true for line coordinates.

Let us now turn to Figure 1.4 for an illustration of (1.1). The line $\underline{\mathbf{L}}$ is represented by the affine vector $\mathbf{l}$, which is the normal vector of the plane containing $\underline{\mathbf{L}}$ and the origin. This plane also contains the affine vector $\mathbf{p}$ from the origin to $\underline{\mathbf{p}}$. This vector is thus orthogonal to the normal vector, as stated in (1.1).

1.3 Points and Lines

Intuitively, two distinct points define a line through them. If our definition of the projective plane is meaningful, then it should support this notion. It does indeed do that: let $\underline{\mathbf{a}}$ and $\underline{\mathbf{b}}$ be two distinct points in $\mathbb{P}^2$. Can we find a line that passes through them? If that line is denoted by $\underline{\mathbf{L}}$, then $\underline{\mathbf{a}}$ lies on it:

$$\underline{\mathbf{La}} = 0,$$

and so does $\underline{\mathbf{b}}$:

$$\underline{\mathbf{Lb}} = 0.$$

It now takes nothing but elementary linear algebra to conclude that

$$\underline{\mathbf{L}} = \underline{\mathbf{a}} \wedge \underline{\mathbf{b}}, \tag{1.2}$$

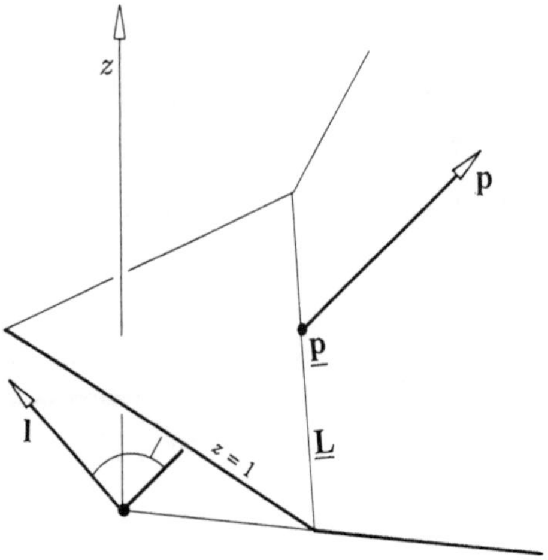

Figure 1.4. A point on a line: If $\underline{\mathbf{p}}$ is on $\underline{\mathbf{L}}$, then $\mathbf{p}$ must be orthogonal to $\mathbf{l}$.

with " $\wedge$ " denoting the cross or vector product. It is computed using the following determinants:

$$\underline{\mathbf{L}} = \left[\begin{vmatrix} a_2 & b_2 \\ a_3 & b_3 \end{vmatrix}, \; - \begin{vmatrix} a_1 & b_1 \\ a_3 & b_3 \end{vmatrix}, \; \begin{vmatrix} a_1 & b_1 \\ a_2 & b_2 \end{vmatrix} \right]. \tag{1.3}$$

If we replace $\underline{\mathbf{a}}$ by a nonzero multiple $\alpha\underline{\mathbf{a}}$, we have not changed $\underline{\mathbf{a}}$'s location. Consequently, $\alpha\underline{\mathbf{a}}$ and $\underline{\mathbf{b}}$ should define the same line as did $\underline{\mathbf{a}}$ and $\underline{\mathbf{b}}$. And they do: $\alpha\underline{\mathbf{a}} \wedge \underline{\mathbf{b}} = \alpha(\underline{\mathbf{a}} \wedge \underline{\mathbf{b}}) = \alpha\underline{\mathbf{L}} \hat{=} \underline{\mathbf{L}}$.

Now for the corresponding concept for lines: the intersection of two lines should define a point. So let $\underline{\mathbf{L}}$ and $\underline{\mathbf{M}}$ denote two lines. If they intersect in a point $\underline{\mathbf{x}}$, then $\underline{\mathbf{x}}$ must satisfy

$$\underline{\mathbf{L}\mathbf{x}} = 0$$

and also

$$\underline{\mathbf{M}\mathbf{x}} = 0.$$

Hence $\underline{\mathbf{x}}$ can be found as a cross product:

$$\underline{\mathbf{x}} = \underline{\mathbf{L}} \wedge \underline{\mathbf{M}}, \tag{1.4}$$

which is quite similar to (1.2). For some examples, see Figure 1.7.

Any two distinct points define a line — that seems fairly obvious. But if any two distinct lines intersect in a point, are we not excluding parallel lines? We are indeed: there are no parallel lines in projective geometry!

1.4 Projective Line Coordinates

Two points define a line — but surely there must be more than two points on a line. If $\underline{\mathbf{a}}$ and $\underline{\mathbf{b}}$ define a line, then what does it take for a third point $\underline{\mathbf{x}}$ to be on that line? Since the line is given by $\underline{\mathbf{a}} \wedge \underline{\mathbf{b}}$, the point $\underline{\mathbf{x}}$ must satisfy

$$[\underline{\mathbf{a}} \wedge \underline{\mathbf{b}}]\underline{\mathbf{x}} = 0,$$

which is equivalent to

$$\det[\underline{\mathbf{a}}, \underline{\mathbf{x}}, \underline{\mathbf{b}}] = 0. \tag{1.5}$$

Equation (1.5) gives a necessary and sufficient condition for three distinct points to lie on the same line, or to be *collinear.*

There is a similar condition for three lines going through one point. Three such lines are called *concurrent.* Using the same approach as for (1.5), we obtain:

Three distinct lines $\underline{\mathbf{L}}, \underline{\mathbf{X}}, \underline{\mathbf{M}}$ are concurrent if and only if

$$\det[\underline{\mathbf{L}}, \underline{\mathbf{X}}, \underline{\mathbf{M}}] = 0. \tag{1.6}$$

If three points are collinear, then their coordinate columns are linearly dependent, as expressed by (1.5). We may thus write one point as a linear combination of the other two:

$$\underline{\mathbf{x}} = \alpha\underline{\mathbf{a}} + \beta\underline{\mathbf{b}}, \tag{1.7}$$

where α and β are suitable real numbers. By letting α and β trace out all reals, $\underline{\mathbf{x}}$ will assume all point locations on the line $\underline{\mathbf{L}}$ through $\underline{\mathbf{a}}$ and $\underline{\mathbf{b}}$. Thus each point on the line is determined by two numbers α and β or any multiple thereof. The pair $[\alpha, \beta]$ is referred to as *projective coordinates* of the line.

We can also make $\underline{\mathbf{x}}$ trace out $\underline{\mathbf{L}}$ by letting it depend on one parameter only. Let us set

$$\underline{\mathbf{x}} = \underline{\mathbf{a}} + t\underline{\mathbf{b}}, \tag{1.8}$$

again with t being any real number. Figure 1.5 illustrates *how* $\underline{\mathbf{x}}$ traces out the line $\underline{\mathbf{L}}$. Note that the point $\underline{\mathbf{b}}$ corresponds to the parameter values $t = \pm\infty$.

When we refer to a line as the collection of all points on it, we call it a *range.* Equation (1.8) gives an example of a *parametrization* of a range.

The same principles hold true for lines: if $\underline{\mathbf{L}}, \underline{\mathbf{X}}$, and $\underline{\mathbf{M}}$ are concurrent, i.e., all contain a point $\underline{\mathbf{p}}$, then we can write one of them as a linear combination of the others:

$$\underline{\mathbf{X}} = \alpha\underline{\mathbf{L}} + \beta\underline{\mathbf{M}}. \tag{1.9}$$

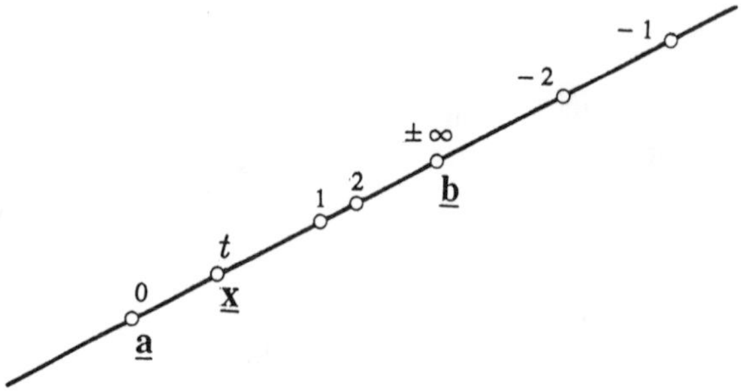

Figure 1.5. Tracing out a line: the point $\underline{\mathbf{x}}$ depends on t, and its location is shown for various values of t.

This will look strange to those who encounter projective geometry for the first time — but clearly a linear combination of two coordinate rows yields another row, and thus denotes a line. Again, we refer to $[\alpha, \beta]$ as *projective coordinates* of the *pencil* of lines through $\underline{\mathbf{p}}$.

As α and β assume all possible real values, (1.9) produces all lines that pass through the intersection of $\underline{\mathbf{L}}$ and $\underline{\mathbf{M}}$. Again, we may wish to express $\underline{\mathbf{X}}$ as depending on one parameter only; then we write

$$\underline{\mathbf{X}} = \underline{\mathbf{L}} + t\underline{\mathbf{M}}. \tag{1.10}$$

Equation (1.10) gives an example of a *parametrization* of a pencil. Figure 1.6 illustrates.

The following "trick" will be helpful in the future. Let $\underline{\mathbf{a}}, \underline{\mathbf{x}}, \underline{\mathbf{b}}$ be three distinct collinear points, such that (1.7) holds. We may scale the components of each point (thus not changing locations!) in a suitable way, so that we can write

$$\underline{\mathbf{x}} \hat{=} \underline{\mathbf{a}} + \underline{\mathbf{b}}. \tag{1.11}$$

Thus if we are given any three distinct collinear points, they may always be rewritten such that they are as in relation (1.11). Example 1.1 gives a numerical demonstration. In terms of projective coordinates on the line, the three points $\underline{\mathbf{a}}, \underline{\mathbf{x}}, \underline{\mathbf{b}}$ may be denoted by $[1, 0], [1, 1], [0, 1]$, respectively. This works only once: if we have assigned these projective coordinates to three points on a line, then the coordinates of all other points are fixed! That is why we say that three distinct points on a line constitute a *projective reference frame* for a line.

Let $\underline{\mathbf{a}}_1, \underline{\mathbf{a}}_2, \underline{\mathbf{a}}_3$ and $\underline{\mathbf{b}}_1, \underline{\mathbf{b}}_2, \underline{\mathbf{b}}_3$ be two projective reference frames for some line $\underline{\mathbf{L}}$. We may always set $\underline{\mathbf{a}}_2 \hat{=} \underline{\mathbf{a}}_1 + \underline{\mathbf{a}}_3$ by properly rescaling all coordinates

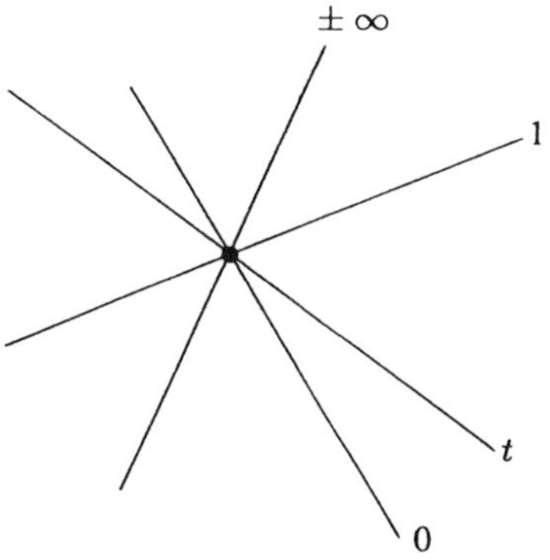

Figure 1.6. Tracing out a pencil: as t traces out all reals, the corresponding line assumes all positions in the pencil.

Let three collinear points be given by

$$\underline{\mathbf{a}} = \begin{bmatrix} 1 \\ 0 \\ 1 \end{bmatrix}, \underline{\mathbf{x}} = \begin{bmatrix} 8 \\ 0 \\ 5 \end{bmatrix}, \underline{\mathbf{b}} = \begin{bmatrix} 2 \\ 0 \\ 1 \end{bmatrix},$$

such that $\underline{\mathbf{x}} = 2\underline{\mathbf{a}} + 3\underline{\mathbf{b}}$. Now replace $\underline{\mathbf{a}}$ by $2\underline{\mathbf{a}}$ and $\underline{\mathbf{b}}$ by $3\underline{\mathbf{b}}$. With these new components, we have $\underline{\mathbf{x}} = \underline{\mathbf{a}} + \underline{\mathbf{b}}$.

Example 1.1. Adjusting point components.

on the line, and in general, we will then not have that $\underline{\mathbf{b}}_2 \hat{=} \underline{\mathbf{b}}_1 + \underline{\mathbf{b}}_3$. But if this *is* the case, we call the two frames *equivalent.* We then have $\underline{\mathbf{a}}_3 - \underline{\mathbf{a}}_1 \hat{=} \underline{\mathbf{b}}_3 - \underline{\mathbf{b}}_1$. This means that both reference frames produce the same parametrization of the line.

The reader is invited to verify that the corresponding statements for lines and pencils also hold. Several points and lines, together with their coordinates, are shown in Figure 1.7.

1.5 Projective Plane Coordinates

Another interesting configuration is given by considering three noncollinear points $\underline{\mathbf{a}}, \underline{\mathbf{b}}, \underline{\mathbf{c}}$ forming a triangle and a fourth point $\underline{\mathbf{d}}$. We can always

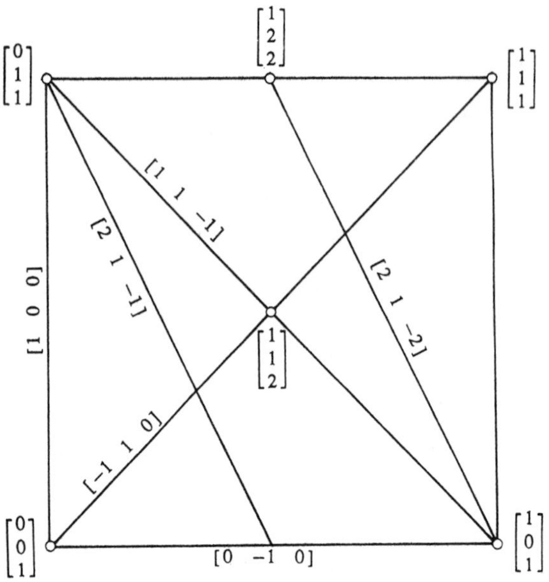

Figure 1.7. Projective coordinates: some examples of points and lines.

write $\underline{\mathbf{d}}$ as

$$\underline{\mathbf{d}} \hat{=} \underline{\mathbf{a}} + \underline{\mathbf{b}} + \underline{\mathbf{c}}. \tag{1.12}$$

To justify this claim, we observe that the point $[\underline{\mathbf{c}} \wedge \underline{\mathbf{d}}] \wedge [\underline{\mathbf{a}} \wedge \underline{\mathbf{b}}]$ can be written as $\underline{\mathbf{a}} + \underline{\mathbf{b}}$, by our above considerations. Thus the labels in Figure 1.8 are justified. We know that $\underline{\mathbf{d}}$ is on the line through $\underline{\mathbf{a}}$ and $\underline{\mathbf{b}} + \underline{\mathbf{c}}$, and also on the line containing $\underline{\mathbf{b}}$ and $\underline{\mathbf{a}} + \underline{\mathbf{c}}$. Thus:

$$\underline{\mathbf{d}} = \lambda \underline{\mathbf{b}} + \underline{\mathbf{a}} + \underline{\mathbf{c}} = \mu \underline{\mathbf{a}} + \underline{\mathbf{b}} + \underline{\mathbf{c}},$$

from which we deduce $\mu = \lambda$, and (1.12) is proved. Note that $\underline{\mathbf{d}}$ can be anywhere, as long as it is not on one of the triangle edges. In particular, $\underline{\mathbf{d}}$ does not have to be "inside" the triangle.[2]

Once four points, no three of them collinear, have been determined, and coordinates have been assigned to them, the coordinates of every point in the projective plane are determined: we say that the four points constitute a *projective reference frame* for $\mathbb{P}^2$. See also the Problems section. Figure 1.9 illustrates the structure of the corresponding coordinate system. There are three centers of pencils corresponding to the coordinates

[2] See Section 1.10 for more details on this topological aspect of the projective plane.

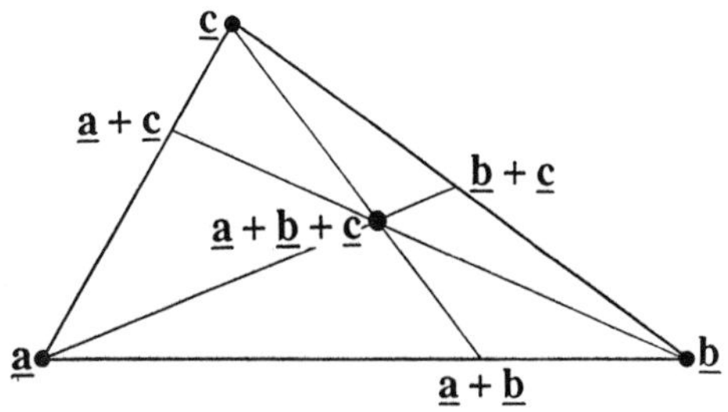

Figure 1.8. Triangle configurations: this labeling of four arbitrary points can always be achieved.

$\underline{\mathbf{f}} = [-\infty, \infty, 0]^T, \underline{\mathbf{g}} = [-\infty, 0, \infty]^T$, and $\underline{\mathbf{h}} = [0, -\infty, \infty]^T$. It is easy to see that they are collinear; we call the line that contains them the *fundamental line* of our coordinate system. It will be used in the treatment of quadrics in Section 13.3.

If we draw the fourth point $\underline{\mathbf{d}}$ as being "outside" the triangle spanned by $\underline{\mathbf{a}}, \underline{\mathbf{b}}, \underline{\mathbf{c}}$, then we obtain a network as shown in Figure 1.10. This network is called a *Moebius net*; see [25]. We also make the stipulation that $\underline{\mathbf{b}} = \underline{\mathbf{a}} + \underline{\mathbf{f}}$ and that $\underline{\mathbf{c}} = \underline{\mathbf{a}} + \underline{\mathbf{g}}$, where $\underline{\mathbf{f}}$ and $\underline{\mathbf{g}}$ are defined as shown in Figure 1.10. We call the line $\underline{\mathbf{F}} = \underline{\mathbf{f}} \wedge \underline{\mathbf{g}}$ the *fundamental line* of our Moebius net.

Then every point $\underline{\mathbf{x}}$ is defined by two coordinates u, v:

$$\underline{\mathbf{x}} = \left[[\underline{\mathbf{a}} + u\underline{\mathbf{g}}] \wedge \underline{\mathbf{f}}\right] \wedge \left[\underline{\mathbf{a}} + v\underline{\mathbf{f}}] \wedge \underline{\mathbf{g}}\right]. \tag{1.13}$$

In particular, $\underline{\mathbf{b}}$ corresponds to $(0,1)$ whereas $\underline{\mathbf{d}}$ corresponds to $(1,1)$.

1.6 Ranges and Pencils

There is an interesting interplay between range and pencil configurations: let $\underline{\mathbf{a}}, \underline{\mathbf{x}}, \underline{\mathbf{b}}$ be three collinear points such that

$$\underline{\mathbf{x}} = u\underline{\mathbf{a}} + v\underline{\mathbf{b}}.$$

Let $\underline{\mathbf{c}}$ be a point not on the line through $\underline{\mathbf{a}}$ and $\underline{\mathbf{b}}$ and define three lines $\underline{\mathbf{A}}, \underline{\mathbf{X}}, \underline{\mathbf{B}}$ by setting $\underline{\mathbf{A}} = \underline{\mathbf{a}} \wedge \underline{\mathbf{c}}$, etc. Then the lines satisfy the same relationship as the points do, i.e.,

$$\underline{\mathbf{X}} = u\underline{\mathbf{A}} + v\underline{\mathbf{B}}. \tag{1.14}$$

For a proof, we refer to Figure 1.11 and write out $\underline{\mathbf{X}}$'s definition:

$$\underline{\mathbf{X}} \quad = \quad [u\underline{\mathbf{a}} + v\underline{\mathbf{b}}] \wedge \underline{\mathbf{c}}$$

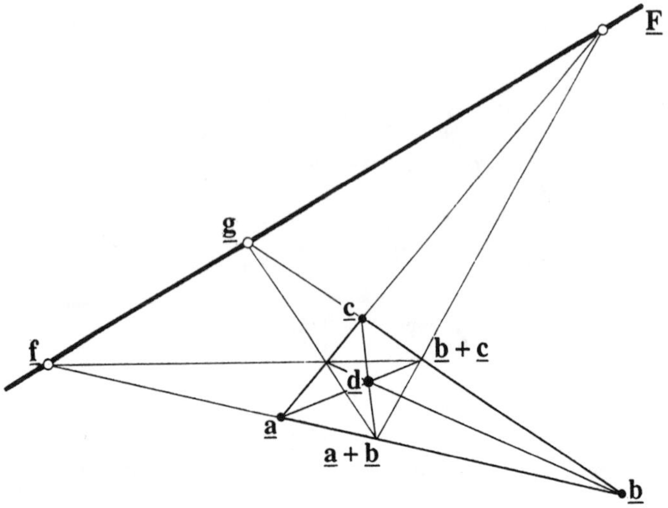

Figure 1.9. Projective coordinate systems: four points form a projective reference frame and determine a fundamental line.

$$
\begin{aligned}
&= u(\underline{\mathbf{a}} \wedge \underline{\mathbf{c}}) + v(\underline{\mathbf{b}} \wedge \underline{\mathbf{c}}) \\
&= u\underline{\mathbf{A}} + v\underline{\mathbf{B}},
\end{aligned}
$$

thus asserting (1.14).

1.7 Pappus' Theorem

This is one of the most fundamental theorems on the projective plane, and will be used in many constructions later on. It can be stated as follows:

Let $\underline{\mathbf{L}}$ and $\underline{\mathbf{M}}$ be two distinct lines, let $\underline{\mathbf{a}}_1, \underline{\mathbf{a}}_2, \underline{\mathbf{a}}_3$ be three distinct points on $\underline{\mathbf{L}}$, and let $\underline{\mathbf{b}}_1, \underline{\mathbf{b}}_2, \underline{\mathbf{b}}_3$ be three distinct points on $\underline{\mathbf{M}}$. Then the three points

$$
\begin{aligned}
\underline{\mathbf{p}}_1 &= [\underline{\mathbf{a}}_2 \wedge \underline{\mathbf{b}}_3] \wedge [\underline{\mathbf{a}}_3 \wedge \underline{\mathbf{b}}_2], \\
\underline{\mathbf{p}}_2 &= [\underline{\mathbf{a}}_1 \wedge \underline{\mathbf{b}}_3] \wedge [\underline{\mathbf{a}}_3 \wedge \underline{\mathbf{b}}_1], \\
\underline{\mathbf{p}}_3 &= [\underline{\mathbf{a}}_1 \wedge \underline{\mathbf{b}}_2] \wedge [\underline{\mathbf{a}}_2 \wedge \underline{\mathbf{b}}_1]
\end{aligned}
\qquad (1.15)
$$

are collinear, as illustrated in Figure 1.12. The line on which they all lie on is called the *Pappus line.*

For a proof of this theorem, assume that $\underline{\mathbf{q}}$ is the intersection of the two lines $\underline{\mathbf{L}}$ and $\underline{\mathbf{M}}$. If necessary, we can rescale the coordinates of all involved

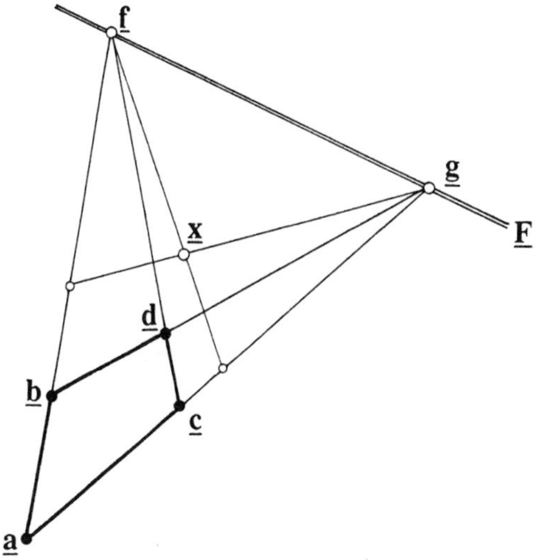

Figure 1.10. Projective coordinate systems: four points may be used to generate a quadrilateral coordinate system.

points such that we have four points on $\underline{\mathbf{L}}$ as follows:

$$\underline{\mathbf{q}},\ \underline{\mathbf{a}}_1,\ \underline{\mathbf{a}}_2 \hat{=} \underline{\mathbf{q}} + \underline{\mathbf{a}}_1,\ \underline{\mathbf{a}}_3 \hat{=} \underline{\mathbf{q}} + \lambda\underline{\mathbf{a}}_1$$

and four points on $\underline{\mathbf{M}}$:

$$\underline{\mathbf{q}},\ \underline{\mathbf{b}}_1 \hat{=} \underline{\mathbf{q}} + \mu\underline{\mathbf{b}}_3,\ \underline{\mathbf{b}}_2 \hat{=} \underline{\mathbf{q}} + \underline{\mathbf{b}}_3,\ \underline{\mathbf{b}}_3.$$

Example 1.1 shows how to achieve this.

Let us now obtain analytic expressions for the points $\underline{\mathbf{p}}_i$. Since $\underline{\mathbf{p}}_1$ is the intersection of the two lines $\underline{\mathbf{a}}_2 \wedge \underline{\mathbf{b}}_3$ and $\underline{\mathbf{a}}_3 \wedge \underline{\mathbf{b}}_2$, we can write

$$\underline{\mathbf{p}}_1 = r(\underline{\mathbf{q}} + \underline{\mathbf{a}}_1) + s\underline{\mathbf{b}}_3 = t(\underline{\mathbf{q}} + \underline{\mathbf{b}}_3) + u(\underline{\mathbf{q}} + \lambda\underline{\mathbf{a}}_1).$$

Since the points $\underline{\mathbf{q}}, \underline{\mathbf{a}}_1, \underline{\mathbf{b}}_3$ are assumed to be noncollinear, we can compare coefficients:

$$\begin{aligned} r &= t + u, \\ r &= u\lambda, \\ s &= t. \end{aligned}$$

A common factor in the r, s, t, u does not matter, so we set $u = 1$ and obtain $r = \lambda$ and $s = t = \lambda - 1$. Thus

$$\underline{\mathbf{p}}_1 = \lambda\underline{\mathbf{q}} + \lambda\underline{\mathbf{a}}_1 + (\lambda - 1)\underline{\mathbf{b}}_3.$$

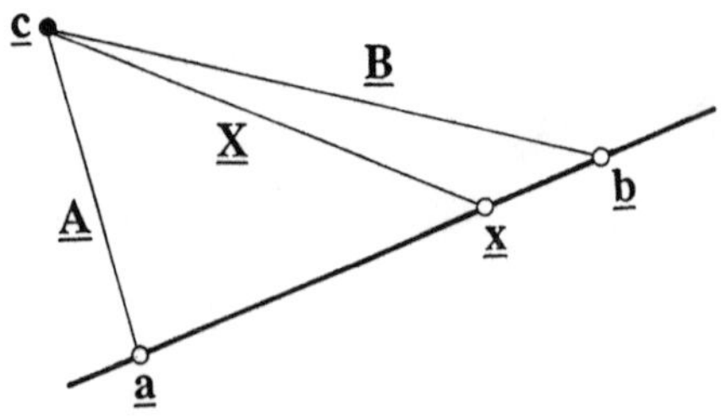

Figure 1.11. Ranges and pencils: collinear points satisfy the same relation as concurrent lines through them.

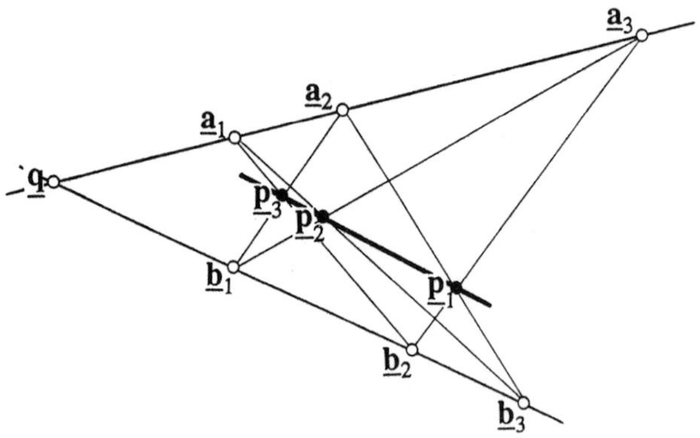

Figure 1.12. Pappus' theorem: the indicated (solid) points are collinear.

Using the same technique, we also find

$$\underline{\mathbf{p}}_2 = \lambda \underline{\mathbf{a}}_1 - \mu \underline{\mathbf{b}}_3$$

and

$$\underline{\mathbf{p}}_3 = \mu \underline{\mathbf{q}} + (\mu - 1)\underline{\mathbf{a}}_1 + \mu \underline{\mathbf{b}}_3.$$

In order to show that $\underline{\mathbf{p}}_1, \underline{\mathbf{p}}_2, \underline{\mathbf{p}}_3$ are indeed collinear, we combine their representations:

$$\begin{bmatrix} \underline{\mathbf{p}}_1 \\ \underline{\mathbf{p}}_2 \\ \underline{\mathbf{p}}_3 \end{bmatrix} = \begin{bmatrix} \lambda & \lambda & \lambda - 1 \\ 0 & \lambda & -\mu \\ \mu & \mu - 1 & \mu \end{bmatrix} \begin{bmatrix} \underline{\mathbf{q}} \\ \underline{\mathbf{a}}_1 \\ \underline{\mathbf{b}}_3 \end{bmatrix}.$$

The determinant of the 3×3 matrix is zero, and so Pappus' theorem is proved.

1.8 Duality

Assume we are given two triples of reals, say a, b, c and r, s, t. If we know that their dot product vanishes, i.e., $ar + bs + ct = 0$, how can we interpret this in terms of projective geometry? We could say that the triple a, b, c represents a point, the triple r, s, t represents a line, and the point lies on the line. Just as well, of course, we might assume that a, b, c is the line and r, s, t is a point on it.

Or consider the statement that one triple of reals is the cross product of two other triples. This could mean that a line is being defined by two points, but it might also mean that a point is being defined as the intersection of two lines.

This possibility of dual geometric interpretations of an algebraic statement is fundamental to all of projective geometry, and is called *the principle of duality*. Algebraically, all statements in projective geometry are concerned with certain determinants or dot products vanishing, or certain cross product identities. All those algebraic statements have two geometric interpretations! The principle of duality goes back to J. Poncelet, J. Gergonne, and J. Plücker and was developed around 1830.

Instead of elaborating on this theoretically, we will produce a dual interpretation of Pappus' theorem. Stated algebraically, Pappus' theorem says: if $\det[\underline{\mathbf{a}}_1, \underline{\mathbf{a}}_2, \underline{\mathbf{a}}_3] = 0$ and $\det[\underline{\mathbf{b}}_1, \underline{\mathbf{b}}_2, \underline{\mathbf{b}}_3] = 0$, then also $\det[\underline{\mathbf{p}}_1, \underline{\mathbf{p}}_2, \underline{\mathbf{p}}_3] = 0$, with the $\underline{\mathbf{p}}_i$ being defined by (1.15).

To obtain the dual statement, we now say: if $\det[\underline{\mathbf{A}}_1, \underline{\mathbf{A}}_2, \underline{\mathbf{A}}_3] = 0$ and $\det[\underline{\mathbf{B}}_1, \underline{\mathbf{B}}_2, \underline{\mathbf{B}}_3] = 0$, then also $\det[\underline{\mathbf{P}}_1, \underline{\mathbf{P}}_2, \underline{\mathbf{P}}_3] = 0$, with $\underline{\mathbf{P}}_1 = [\underline{\mathbf{A}}_2 \wedge \underline{\mathbf{B}}_3] \wedge [\underline{\mathbf{A}}_3 \wedge \underline{\mathbf{B}}_2]$, etc. Since we already proved Pappus' theorem, this dual version comes for free! Its proof would be a virtual carbon copy of the proof of Pappus' theorem, and can therefore be omitted.

We still need a geometric interpretation of this dual Pappus theorem: it is a statement about two triples of concurrent lines. Intersect corresponding lines as shown in Figure 1.13, and form lines $\underline{\mathbf{P}}_i$ as shown. The dual Pappus theorem states that all three $\underline{\mathbf{P}}_i$ are concurrent. The point in which they all meet in is called the *Pappus point.*

1.9 The Extended Affine Plane

We now return to the discussion from the beginning of this chapter. There we considered the plane $z = 1$ in Euclidean three-space and projected 3-D lines and planes into it, thus obtaining points and lines of the projective plane $\mathbb{P}^2$. Taken simply as a subset of three-space, the plane $z = 1$ is an affine plane. We shall now discuss these two interpretations of the plane

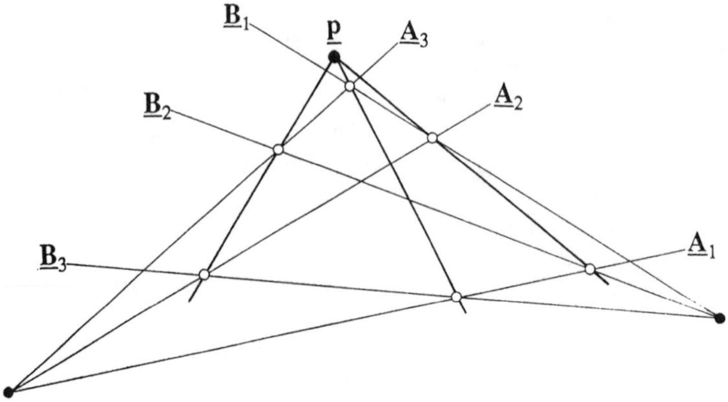

Figure 1.13. The dual Pappus theorem: the indicated (heavy) lines are concurrent.

— projective and affine.

A point $\mathbf{x} = [x, y]^{\mathrm{T}}$ of the affine plane may be identified with the 3-D point $\underline{\mathbf{x}} = [x, y, 1]^{\mathrm{T}}$ in the plane $z = 1$. Interpreted as a projective point, $\underline{\mathbf{x}}$ is also identified with all other points $\underline{\mathbf{x}} \hat{=} [\alpha x, \alpha y, \alpha]^{\mathrm{T}}$ for $\alpha \neq 0$. See Figure 1.2 for an illustration.

We then have: a projective point $\underline{\mathbf{x}} = [x, y, z]^{\mathrm{T}}$ corresponds to the affine point $\mathbf{x} = [x/z, y/z]^{\mathrm{T}}$. We say that $\underline{\mathbf{x}}$ are the *homogeneous coordinates* of $\mathbf{x}$. For now we must assume $z \neq 0$; we will treat the case $z = 0$ shortly.

What is the affine counterpart $\mathbf{L}$ of a projective line $\underline{\mathbf{L}}$? If $\underline{\mathbf{L}} = [l, m, n]$, then $\mathbf{L} = [l/n, m/n]$. This has to be interpreted as the coefficients of $\mathbf{L}$'s implicit equation $(l/n)x + (m/n)y + 1 = 0$. Again, we must require that $n \neq 0$.

If

$$\underline{\mathbf{x}} = \alpha\underline{\mathbf{a}} + \beta\underline{\mathbf{b}} = \begin{bmatrix} \alpha a_x + \beta b_x \\ \alpha a_y + \beta b_y \\ \alpha a_z + \beta b_z \end{bmatrix}$$

holds for three projective points, what is the corresponding relation for their affine counterparts? We obtain

$$\begin{aligned} \mathbf{x} &= \frac{1}{\alpha a_z + \beta b_z} \begin{bmatrix} \alpha a_x + \beta b_x \\ \alpha a_y + \beta b_y \end{bmatrix} \\ &= \frac{\alpha a_z}{\alpha a_z + \beta b_z} \begin{bmatrix} a_x/a_z \\ a_y/a_z \end{bmatrix} + \frac{\beta b_z}{\alpha a_z + \beta b_z} \begin{bmatrix} b_x/b_z \\ b_y/b_z \end{bmatrix} \end{aligned}$$

$$= \frac{\alpha a_z}{\alpha a_z + \beta b_z}\mathbf{a} + \frac{\beta b_z}{\alpha a_z + \beta b_z}\mathbf{b}.$$

Notice how the coefficients of the affine points $\mathbf{a}$ and $\mathbf{b}$ sum to one, thus forming a *barycentric combination*, also called an affine combination. While we can form arbitrary linear combinations of projective points, this is illegal for affine points; the only admissable combination for them is the barycentric combination.

In affine space, we can talk about phenomena that do not exist in projective space. For example: affine lines may be parallel, while projective ones cannot. Take two projective lines $\underline{\mathbf{L}} = [l, m, n]$ and $\underline{\mathbf{M}} = [l, m, k]$. They intersect in the projective point $\underline{\mathbf{L}} \wedge \underline{\mathbf{M}}$, whose third component is zero. Thus it does not have an affine counterpart! We remedy this situation by adding *points at infinity* to the set of affine points. These are all the points with zero third components.

The name "points at infinity" is justified by the informal observation that the intersection of two lines moves further and further towards infinity as the lines become more and more parallel. All affine points at infinity lie on a projective line: it is the line $\underline{\mathbf{L}} = [0, 0, 1]$. Its affine counterpart is the *line at infinity* $\mathbf{L}_\infty$. If we "add" this line to the affine plane, we speak of the extended affine plane $\mathbb{E}^2$. We may think of this line as the "horizon" of the affine plane. Another word for points at infinity is "ideal points," lying on the "ideal line." The extended affine plane is a model for the projective plane in the same sense as the models that will be discussed in Section 1.10.

The difference $\mathbf{a} - \mathbf{b}$ of two affine points is a *vector*. But the difference $\underline{\mathbf{a}} - \underline{\mathbf{b}}$ of two projective points is yet another projective point; it is on the line through $\underline{\mathbf{a}}$ and $\underline{\mathbf{b}}$. Let $\underline{\mathbf{a}} = [a_x, a_y, 1]^{\mathrm{T}}$ and $\underline{\mathbf{b}} = [b_x, b_y, 1]^{\mathrm{T}}$ be the projective versions of two affine points $\mathbf{a}$ and $\mathbf{b}$. The difference of the affine points is an affine vector; the difference of the projective points corresponds to an affine point at infinity. Thus affine points at infinity correspond to vectors!

The affine vector $\mathbf{x}$ corresponds to a point at infinity, but so does $\alpha\mathbf{x}$ for any $\alpha \neq 0$. Thus a point at infinity does not only correspond to a single vector, but to all vectors "pointing" to it.

1.10 Topological Aspects

In our investigation of the projective plane, we have so far pointed out *geometric* features, focusing on properties such as collinearity, duality, etc. There is another aspect that we have not addressed yet: the projective plane has a different *topology* than the affine plane. The realization of this fact led to some investigations by computer graphics researchers, since homogeneous coordinates (and hence projective geometry) are one of the

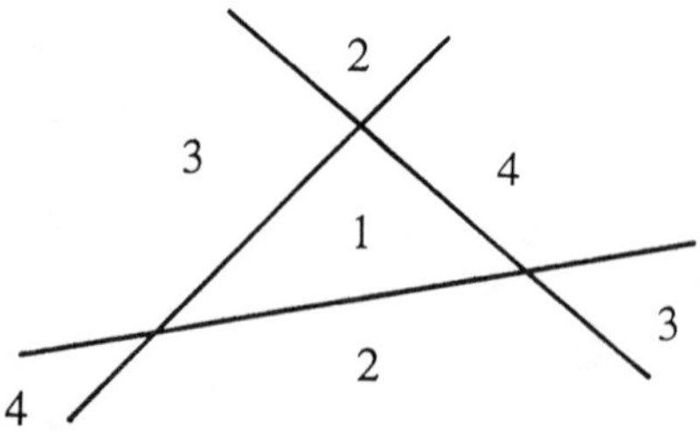

Figure 1.14. A division of the projective plane: three lines divide it into four regions.

standard tools in that field. See Riesenfeld [99] for an introduction.

Let $\underline{\mathbf{L}}, \underline{\mathbf{M}}, \underline{\mathbf{N}}$ be three nonconcurrent lines in the projective plane. They divide $\mathbb{P}^2$ into four regions: let $\underline{\mathbf{a}}, \underline{\mathbf{b}}, \underline{\mathbf{c}}$ be their intersections, and also pick a fourth point $\underline{\mathbf{d}}$ not on any of the lines. These four points form a projective reference frame, as illustrated by Figure 1.8. We can identify the following four regions:

$$\mathcal{R}_1 : \begin{bmatrix} + \\ + \\ + \end{bmatrix}, \quad \mathcal{R}_2 : \begin{bmatrix} - \\ + \\ + \end{bmatrix}, \quad \mathcal{R}_3 : \begin{bmatrix} + \\ - \\ + \end{bmatrix}, \quad \mathcal{R}_4 : \begin{bmatrix} + \\ + \\ - \end{bmatrix},$$

where "+" stands for points with positive projective coordinates, and "-" for those with negative ones. Figure 1.14 illustrates.

This property may be utilized to derive what is known as the *circular model* of $\mathbb{P}^2$: referring to Figure 1.15, we identify all *points* inside the unit disc with points in the projective plane. The *lines* in our model are all those circles that intersect the unit circle in two antipodal points; we treat these as being identical; they correspond to the same point at infinity. The unit circle itself then corresponds to the line at infinity.

In order to make two antipodal points on the unit circle coincide in a "physical" model of $\mathbb{P}^2$, we could lift our circular disc out of the plane, deform it, and "glue" corresponding antipodal points together. Since we must do this for *all* point pairs of the unit circle, we will obtain a self-intersecting surface. The book by Hilbert and Cohn-Vossen [66] gives a beautiful illustration.

A different way of deforming the circular model results in *Boy's surface.* A steel model of this intriguing surface has been erected[3] in front of the Mathematics Research Center in Oberwolfach, Germany.

[3]The structure was designed and manufactured by Mercedes-Benz, using a bicubic B-spline approximation.

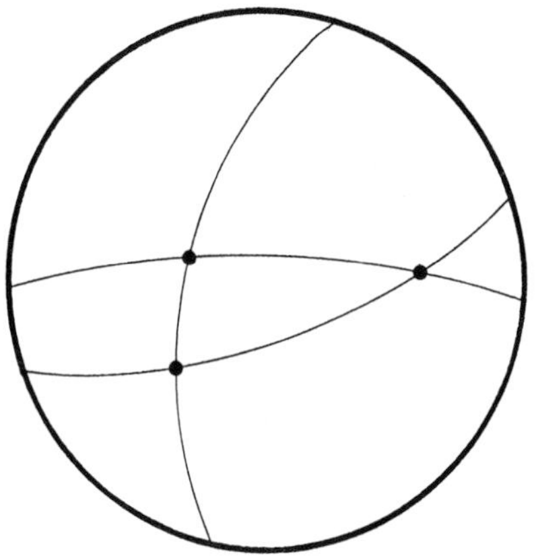

Figure 1.15. The circular model of the projective plane: the unit disc models the projective plane.

A completely different model of the projective plane is the *spherical model.* A point of the projective plane is identified with a pair of antipodal points on the unit sphere. Lines in the projective plane are identified with great circles on the sphere. Any two distinct great circles define a point, and any two distinct points define a great circle.

This model, illustrated by Figure 1.16, shows best that the projective plane has a different topology than the affine plane. Any straight line divides the affine plane into two separate half planes. A point in one half plane can only be connected to one in the other half plane by crossing the line. But a line in the projective plane does not divide it: a great circle does not divide our spherical model. Every point, being identified with its antipodal point, is at the same time on either side of the great circle!

Thus the notion of a half plane is meaningless in projective geometry. Similarly, it is meaningless to talk about a point being "inside" or "outside" a given triangle.[4] It is possible to modify the classical definition of the projective plane so as to allow for the concepts of half planes, or, more generally, of *orientation.* In essence, oriented projective geometry only treats coordinates as being equivalent if they differ by a *positive* factor; see J. Stolfi. [110]

[4]This can cause problems in computer graphics when clipping is performed in homogeneous coordinates; see Blinn/Newell [18].

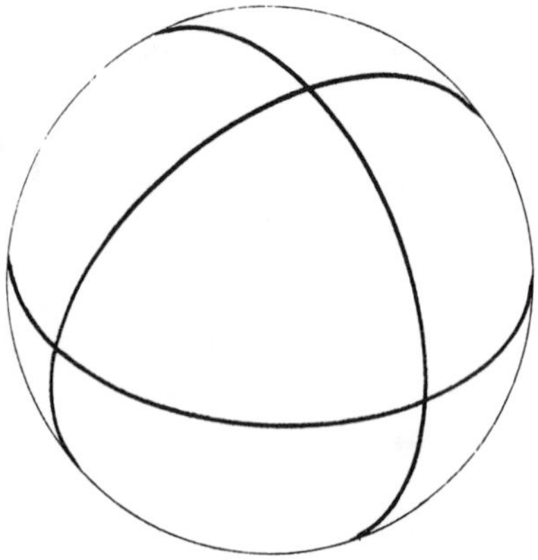

Figure 1.16. The spherical model of the projective plane: projective points are identified with antipodal point pairs; projective lines are identified with great circles.

1.11 Projective Three-space

All projective concepts so far were restricted to the plane. Once the planar case is understood, it is not hard to explore higher-dimensional spaces, in particular projective three-space. Luckily, no unexpected phenomena arise. We now deal with points, lines, and planes as the basic entities. Any two planes intersect in a line, a line and a plane intersect in a point (or the line is contained in the plane), and two lines may intersect (then they are *coplanar*) or not (then they are *skew*). Planes and points are dual to each other, while lines are dual to themselves. If one were to pursue three-dimensional projective geometry in detail, a different set of symbols would be necessary (involving either tensors or so-called Plücker coordinates).

A point $\underline{\mathbf{p}} \in \mathbb{P}^3$ is expressed in terms of four coordinates, not all of them zero:

$$\underline{\mathbf{p}} = \begin{bmatrix} x \\ y \\ z \\ w \end{bmatrix}.$$

Our previous intuitive development interpreted $\mathbb{P}^2$ as the projection of $\mathbb{E}^3$ into the plane $z = 1$ through the origin. We may now think of $\mathbb{P}^3$ as the projection of $\mathbb{E}^4$ into the (three-dimensional) hyperplane $w = 1$ through the origin.

A plane is also represented by a quadruple of reals, written as a row:

$$\underline{\mathbf{P}} = [k, l, m, n].$$

A point $\underline{\mathbf{p}}$ lies on a plane $\underline{\mathbf{P}}$ if $\underline{\mathbf{P}}\underline{\mathbf{p}} = 0$.

Three points $\underline{\mathbf{p}}_1, \underline{\mathbf{p}}_2, \underline{\mathbf{p}}_3$ determine a plane $\underline{\mathbf{P}}$:

$$\underline{\mathbf{P}} = \left[\det P_1, -\det P_2, \det P_3, -\det P_4 \right], \tag{1.16}$$

where P_i is obtained by deleting the ith row from the matrix

$$\begin{bmatrix} x_1 & x_2 & x_3 \\ y_1 & y_2 & y_3 \\ z_1 & z_2 & z_3 \\ w_1 & w_2 & w_3 \end{bmatrix}.$$

We will use the notation

$$\underline{\mathbf{P}} = < \underline{\mathbf{p}}_1, \underline{\mathbf{p}}_2, \underline{\mathbf{p}}_3 > .$$

In the same manner, three planes $\underline{\mathbf{P}}_1, \underline{\mathbf{P}}_2, \underline{\mathbf{P}}_3$ define the point $\underline{\mathbf{p}}$ in which they intersect:

$$\underline{\mathbf{p}} = < \underline{\mathbf{P}}_1, \underline{\mathbf{P}}_2, \underline{\mathbf{P}}_3 > .$$

The line obtained by connecting two points or by intersecting two planes is not so easily described – here, the concept of *Plücker* coordinates is used, but we shall not describe it here. See [111] for details.

1.12 Problems

1. Equation (1.2) shows how to compute the line through two distinct points. Draw a figure like Figure 1.4 to illustrate the idea.

2. After completing Problem 1, illustrate (1.4).

3. Use the spherical model of the projective plane to model the extended affine plane; include the definition of parallel lines.

4. What are the coordinates of the fundamental line from Section 1.5?

5. What is the dual of Figure 1.9?

2

Projective Maps

Geometries are more than just a set of objects, like lines and points. A geometry is also characterized by properties of these objects under the maps of the geometry. Thus Euclidean geometry is characterized by Euclidean maps, which leave lengths and angles unchanged. Affine geometry is characterized by affine maps, which leave the ratio of three collinear points unchanged. Finally, projective geometry is characterized by projective maps which leave the *cross ratio* of four collinear points unchanged. In this chapter, we shall discuss those aspects of projective maps that later will be relevant for NURBS.

2.1 Perspectivities

We have mentioned projections of three-space onto a plane before. At a lower level, one might try to project from a line onto another line, as illustrated in Figure 2.1. Let two points $\underline{\mathbf{p}}_0, \underline{\mathbf{p}}_1$ be given on a line $\underline{\mathbf{L}}$ and two points $\underline{\mathbf{q}}_0, \underline{\mathbf{q}}_1$ on another line $\underline{\mathbf{M}}$. If $\underline{\mathbf{c}}$ is a point on neither $\underline{\mathbf{L}}$ nor $\underline{\mathbf{M}}$, then we may define a projection of $\underline{\mathbf{L}}$ onto $\underline{\mathbf{M}}$ through $\underline{\mathbf{c}}$ in the following way: if $\underline{\mathbf{x}}$ is on $\underline{\mathbf{L}}$, then its image $\underline{\mathbf{y}}$ on $\underline{\mathbf{M}}$ under a projection Φ is defined by

$$\underline{\mathbf{y}} = \Phi(\underline{\mathbf{x}}) = [\underline{\mathbf{c}} \wedge \underline{\mathbf{x}}] \wedge \underline{\mathbf{M}}. \tag{2.1}$$

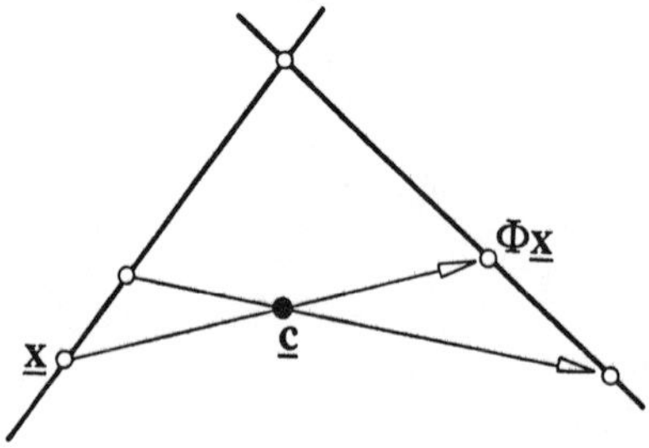

Figure 2.1. Perspectivities: $\underline{\mathbf{y}} = \Phi\underline{\mathbf{x}}$ is defined as a projection through $\underline{\mathbf{c}}$ onto $\underline{\mathbf{M}}$.

Such maps are called *perspectivities*. The point $\underline{\mathbf{c}}$ is called the *center* of the perspectivity. We see immediately that the inverse map from $\underline{\mathbf{M}}$ onto $\underline{\mathbf{L}}$ is defined in the same way, making Φ a bijection.[1]

Let $\underline{\mathbf{a}}$ be the intersection of $\underline{\mathbf{L}}$ and $\underline{\mathbf{M}}$. Then $\Phi\underline{\mathbf{a}} = \underline{\mathbf{a}}$, making $\underline{\mathbf{a}}$ a *fixed point* of the perspectivity. Let

$$\underline{\mathbf{x}} = u\underline{\mathbf{p}}_0 + v\underline{\mathbf{p}}_1$$

and

$$\underline{\mathbf{y}} = \hat{u}\underline{\mathbf{q}}_0 + \hat{v}\underline{\mathbf{q}}_1.$$

Since the points $\underline{\mathbf{c}}, \underline{\mathbf{x}}, \underline{\mathbf{y}}$ are collinear by construction, we have that

$$\det[\underline{\mathbf{c}}, u\underline{\mathbf{p}}_0 + v\underline{\mathbf{p}}_1, \hat{u}\underline{\mathbf{q}}_0 + \hat{v}\underline{\mathbf{q}}_1] = 0.$$

Expanding the determinant, we obtain

$$[u, v]\begin{bmatrix} m_{00} & m_{01} \\ m_{10} & m_{11} \end{bmatrix}\begin{bmatrix} \hat{u} \\ \hat{v} \end{bmatrix} = 0, \tag{2.2}$$

where we have used the abbreviations

$$m_{ij} = \det[\underline{\mathbf{c}}, \underline{\mathbf{p}}_i, \underline{\mathbf{q}}_j].$$

Equation (2.2) relates projective coordinates $\mathbf{u} = [u, v]^{\mathrm{T}}$ on the preimage line to the image coordinates $\hat{\mathbf{u}} = [\hat{u}, \hat{v}]^{\mathrm{T}}$, the image being defined by a perspectivity. Setting $M = \{m_{ij}\}$, we may shorten (2.2) to

$$\mathbf{u}^{\mathrm{T}} M \hat{\mathbf{u}} = 0. \tag{2.3}$$

We note a special case: if $\underline{\mathbf{q}}_0 = \Phi\underline{\mathbf{p}}_0$ and $\underline{\mathbf{q}}_1 = \Phi\underline{\mathbf{p}}_1$, then M's diagonal entries are zero.

---

[1]The reader is invited to work out the dual concept: mapping a pencil onto another pencil!

2.2 Cross Ratios

As mentioned in the introduction to this chapter, cross ratios play a central role in projective geometry. They are defined as follows: Let $\underline{\mathbf{L}}$ be a line and $\underline{\mathbf{a}}$ and $\underline{\mathbf{b}}$ two points on it. If two additional points are defined by

$$\underline{\mathbf{x}} = u_1\underline{\mathbf{a}} + v_1\underline{\mathbf{b}} \quad \text{and} \quad \underline{\mathbf{y}} = u_2\underline{\mathbf{a}} + v_2\underline{\mathbf{b}},$$

then their *cross ratio* cr is given by

$$\text{cr}(\underline{\mathbf{a}}, \underline{\mathbf{x}}, \underline{\mathbf{y}}, \underline{\mathbf{b}}) = \frac{v_1}{u_1}\frac{u_2}{v_2}. \tag{2.4}$$

This particular definition will be useful later; it is not, however, what one finds in classical texts on projective geometry. The commonly used classical definition $\text{cr}_{\text{classic}}$ is related to ours by

$$\text{cr}(\underline{\mathbf{a}}, \underline{\mathbf{x}}, \underline{\mathbf{y}}, \underline{\mathbf{b}}) = \text{cr}_{\text{classic}}(\underline{\mathbf{x}}, \underline{\mathbf{a}}, \underline{\mathbf{y}}, \underline{\mathbf{b}}),$$

and so both definitions can be viewed as being equivalent.

Note that we may multiply the coordinates of any point by an arbitrary nonzero number, and the cross ratio will not change! Also note that $\text{cr}(\underline{\mathbf{a}}, \underline{\mathbf{x}}, \underline{\mathbf{y}}, \underline{\mathbf{b}}) = \text{cr}(\underline{\mathbf{b}}, \underline{\mathbf{y}}, \underline{\mathbf{x}}, \underline{\mathbf{a}})$.

Now let $\underline{\mathbf{c}}$ be a point not on $\underline{\mathbf{L}}$, and define

$$\begin{aligned}
\underline{\mathbf{A}} &= \underline{\mathbf{a}} \wedge \underline{\mathbf{c}}, \\
\underline{\mathbf{X}} &= \underline{\mathbf{x}} \wedge \underline{\mathbf{c}}, \\
\underline{\mathbf{Y}} &= \underline{\mathbf{y}} \wedge \underline{\mathbf{c}}, \\
\underline{\mathbf{B}} &= \underline{\mathbf{b}} \wedge \underline{\mathbf{c}}.
\end{aligned}$$

Since $\underline{\mathbf{X}} = u_1\underline{\mathbf{A}} + v_1\underline{\mathbf{B}}$ and $\underline{\mathbf{Y}} = u_2\underline{\mathbf{A}} + v_2\underline{\mathbf{B}}$ because of (1.14), we can adopt the same definition for the cross ratio of four concurrent lines:

$$\text{cr}(\underline{\mathbf{A}}, \underline{\mathbf{X}}, \underline{\mathbf{Y}}, \underline{\mathbf{B}}) = \text{cr}(\underline{\mathbf{a}}, \underline{\mathbf{x}}, \underline{\mathbf{y}}, \underline{\mathbf{b}}). \tag{2.5}$$

See Figure 2.2 for an illustration.

We may now state that *perspectivities leave cross ratios invariant.* In Figure 2.2, we may draw any other line $\underline{\mathbf{M}}$. It will intersect $\underline{\mathbf{A}}, \underline{\mathbf{X}}, \underline{\mathbf{Y}}, \underline{\mathbf{B}}$ in four points — and those four points, being related to $\underline{\mathbf{a}}, \underline{\mathbf{x}}, \underline{\mathbf{y}}, \underline{\mathbf{b}}$ by a perspectivity, will be also in the same cross ratio as $\underline{\mathbf{A}}, \underline{\mathbf{X}}, \underline{\mathbf{Y}}, \underline{\mathbf{B}}$ and hence as $\underline{\mathbf{a}}, \underline{\mathbf{x}}, \underline{\mathbf{y}}, \underline{\mathbf{b}}$.

A perspectivity from a line $\underline{\mathbf{L}}$ to another line $\underline{\mathbf{M}}$ is determined by two preimage points on $\underline{\mathbf{L}}$ and two image points on $\underline{\mathbf{M}}$, since this is enough information to determine the center $\underline{\mathbf{c}}$.

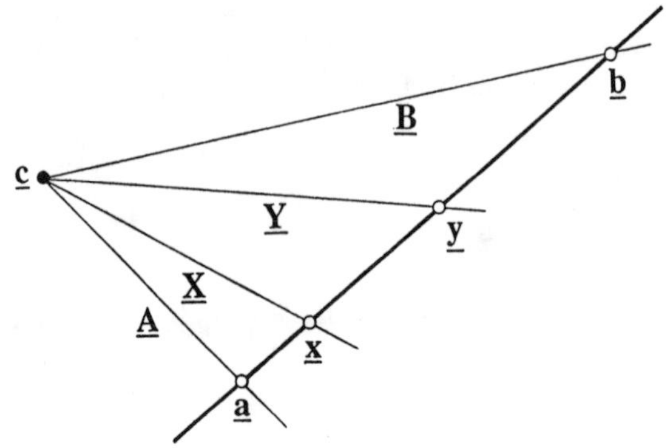

Figure 2.2. Cross ratios: the four collinear points have the same cross ratio as the four concurrent lines.

Four points with cross ratio 1/2 are called *harmonic.*[2] We sometimes relax this definition and label points as harmonic if some permutation of them yields the cross ratio 1/2. Harmonic points will play an important role in the projective theory of conic sections.

If $\underline{\mathbf{a}}$ and $\underline{\mathbf{b}}$ are two distinct points, then

$$\operatorname{cr}(\underline{\mathbf{a}}, \underline{\mathbf{a}}+\underline{\mathbf{b}}, \underline{\mathbf{b}}, \underline{\mathbf{b}}-\underline{\mathbf{a}}) = \frac{1}{2}. \tag{2.6}$$

We see this by writing

$$\begin{aligned} \underline{\mathbf{a}}+\underline{\mathbf{b}} &= 2\cdot\underline{\mathbf{a}} + 1\cdot(\underline{\mathbf{b}}-\underline{\mathbf{a}}), \\ \underline{\mathbf{b}} &= 1\cdot\underline{\mathbf{a}} + 1\cdot(\underline{\mathbf{b}}-\underline{\mathbf{a}}), \end{aligned}$$

from which our claim follows immediately. In an affine setting, this amounts to stating that two points, their midpoint, and their difference vector are harmonic.

Another set of harmonic points is found by adding a line and a point to Figure 1.8, as shown in Figure 2.3. If we draw the line through $\underline{\mathbf{a}}+\underline{\mathbf{c}}$ and $\underline{\mathbf{b}}+\underline{\mathbf{c}}$ and intersect it with $\underline{\mathbf{a}}\wedge\underline{\mathbf{b}}$, the resulting point is $\underline{\mathbf{a}}-\underline{\mathbf{b}}$, which is, of course, equivalent to $\underline{\mathbf{b}}-\underline{\mathbf{a}}$. Again, we have four harmonic points on the line $\underline{\mathbf{a}}\wedge\underline{\mathbf{b}}$.

We could have started this construction by interpreting the four points $\underline{\mathbf{d}}, \underline{\mathbf{b}}+\underline{\mathbf{c}}, \underline{\mathbf{c}}$, and $\underline{\mathbf{a}}+\underline{\mathbf{c}}$ in Figure 2.3 as our initial quadrilateral. Thus four

[2]In the classical definition of cross ratios, they are called harmonic if their "classic cross ratio" $\mathrm{cr}_{\mathrm{classic}}$ equals -1.

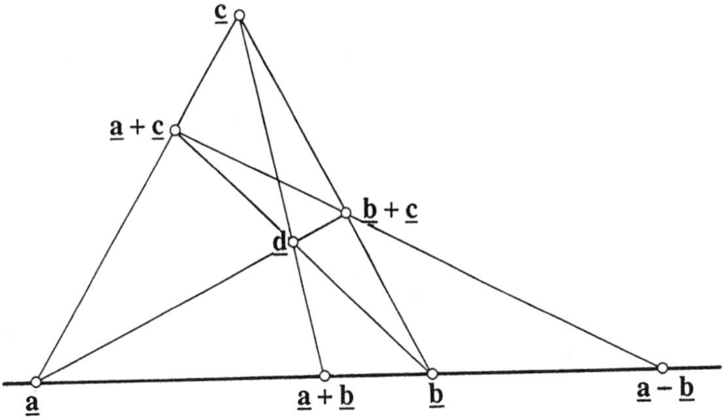

Figure 2.3. Harmonic points: the point $\underline{\mathbf{d}}$ generates four harmonic points on the line $\underline{\mathbf{a}} \wedge \underline{\mathbf{b}}$.

harmonic points are obtained as follows: take any quadrilateral, and intersect opposite edges, resulting in two points. Intersect the line formed by these two points with the two diagonals of the quadrilateral, and we have constructed four harmonic points. A quadrilateral with both diagonals added to it is called a *complete quadrilateral.* Thus complete quadrilaterals are intimately connected with harmonic points.

If three points $\underline{\mathbf{a}}, \underline{\mathbf{b}}, \underline{\mathbf{c}}$ are given on a line, we may ask for the point $\underline{\mathbf{x}}$ that satisfies

$$\operatorname{cr}(\underline{\mathbf{a}}, \underline{\mathbf{x}}, \underline{\mathbf{b}}, \underline{\mathbf{c}}) = \rho$$

for some given cross ratio ρ. This point is uniquely defined. To compute it, assume that $\underline{\mathbf{b}} = u\underline{\mathbf{a}} + v\underline{\mathbf{c}}$. Let us choose the representation $\underline{\mathbf{x}} = \underline{\mathbf{a}} + t\underline{\mathbf{c}}$ for the point $\underline{\mathbf{x}}$. We thus need to find t; it can be computed from the known cross ratio relation

$$\rho = \frac{u}{v} \cdot t. \tag{2.7}$$

2.3 Projectivities

Perspectivities are only a subset of a more general type of map from one line to another. Theire sole characterization is that they leave cross ratios unchanged. Such maps are called *projectivities.*

We will now proceed to give a geometric construction for a projectivity.

Let $\underline{\mathbf{p}}_0, \underline{\mathbf{p}}_1, \underline{\mathbf{p}}_2$ be three points on a line $\underline{\mathbf{L}}$. On a second line $\underline{\mathbf{M}}$, we prescribe three points $\underline{\mathbf{q}}_0, \underline{\mathbf{q}}_1, \underline{\mathbf{q}}_2$ to be the images of the corresponding $\underline{\mathbf{p}}_i$. Our aim is this: for any point $\underline{\mathbf{x}}$ on $\underline{\mathbf{L}}$, find an image point $\hat{\underline{\mathbf{x}}}$ on $\underline{\mathbf{M}}$ such that

$$\mathrm{cr}(\underline{\mathbf{p}}_0, \underline{\mathbf{x}}, \underline{\mathbf{p}}_1, \underline{\mathbf{p}}_2) = \mathrm{cr}(\underline{\mathbf{q}}_0, \hat{\underline{\mathbf{x}}}, \underline{\mathbf{q}}_1, \underline{\mathbf{q}}_2).$$

See Figure 2.4 for an illustration. If we are able to produce an $\hat{\underline{\mathbf{x}}}$ for any $\underline{\mathbf{x}}$, we have defined a projectivity.

We proceed as follows: first, we recall that Pappus' theorem (Section 1.7) asserts the existence of the line $\underline{\mathbf{P}}$, namely the Pappus line, with points $\underline{\mathbf{r}}_0, \underline{\mathbf{r}}_1, \underline{\mathbf{r}}_2$ on it as shown in Figure 2.4.

Next, we define a perspectivity from $\underline{\mathbf{L}}$ to $\underline{\mathbf{P}}$ with center $\underline{\mathbf{q}}_0$. It will map $\underline{\mathbf{x}}$ to a point $\underline{\mathbf{y}}$ on $\underline{\mathbf{P}}$. We observe that

$$\mathrm{cr}(\underline{\mathbf{p}}_0, \underline{\mathbf{x}}, \underline{\mathbf{p}}_1, \underline{\mathbf{p}}_2) = \mathrm{cr}(\underline{\mathbf{b}}, \underline{\mathbf{y}}, \underline{\mathbf{r}}_2, \underline{\mathbf{r}}_1),$$

with $\underline{\mathbf{b}} = [\underline{\mathbf{p}}_0 \wedge \underline{\mathbf{q}}_0] \wedge \underline{\mathbf{P}}$. Then we define another perspectivity; with center $\underline{\mathbf{p}}_0$ and mapping $\underline{\mathbf{P}}$ to $\underline{\mathbf{M}}$. It maps $\underline{\mathbf{y}}$ to the desired point $\hat{\underline{\mathbf{x}}}$ on $\underline{\mathbf{M}}$. We have

$$\mathrm{cr}(\underline{\mathbf{b}}, \underline{\mathbf{y}}, \underline{\mathbf{r}}_2, \underline{\mathbf{r}}_1) = \mathrm{cr}(\underline{\mathbf{q}}_0, \hat{\underline{\mathbf{x}}}, \underline{\mathbf{q}}_1, \underline{\mathbf{q}}_2),$$

thus asserting that we have in fact found a projectivity from $\underline{\mathbf{L}}$ to $\underline{\mathbf{M}}$.

In fact, we have not only shown that a projectivity from one line to another is defined by two triples of images and preimages, but we have also shown that each projectivity may be composed as the concatenation of *two*

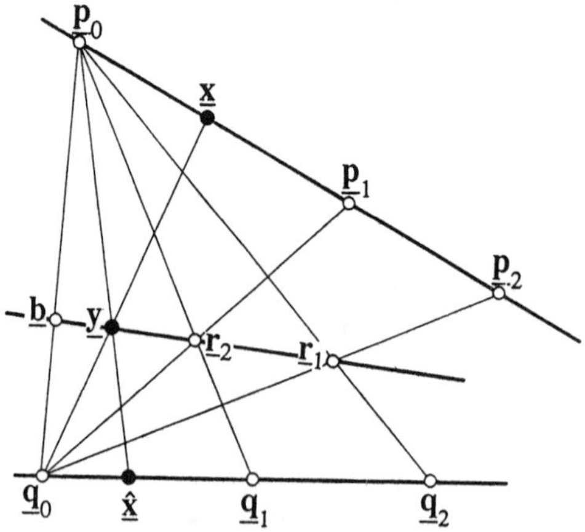

Figure 2.4. Projectivities: $\underline{\mathbf{x}}$'s image $\hat{\underline{\mathbf{x}}}$ may be obtained from two perspectivities.

perspectivities. If the image and preimage lines are identical, we will need three perspectivities.

Our approach to projectivities was a purely constructive one: given three images and preimages, we then constructed an image point for any given preimage point. A more analytical description of projectivities is also available; we will now show that every projectivity can be written in matrix form:

$$\hat{\underline{\mathbf{u}}} = \Phi(\underline{\mathbf{u}}) = M\underline{\mathbf{u}}, \tag{2.8}$$

with $\underline{\mathbf{u}}$ and $\hat{\underline{\mathbf{u}}}$ projective coordinates on the lines $\underline{\mathbf{L}}$ and $\underline{\mathbf{M}}$, respectively.

All we have to show is that the map Φ preserves cross ratios. Let $\underline{\mathbf{p}}_1 = u_1\underline{\mathbf{p}}_0 + v_1\underline{\mathbf{p}}_2$ and $\underline{\mathbf{x}} = u_2\underline{\mathbf{p}}_0 + v_2\underline{\mathbf{p}}_2$. Then

$$\begin{aligned}\hat{\underline{\mathbf{x}}} &= M(u_2\underline{\mathbf{p}}_0 + v_2\underline{\mathbf{p}}_2)\\ &= u_2M\underline{\mathbf{p}}_0 + v_2M\underline{\mathbf{p}}_2\\ &= u_2\underline{\mathbf{q}}_0 + v_2\underline{\mathbf{q}}_2.\end{aligned}$$

The same holds for $\underline{\mathbf{p}}_1$ – thus M leaves linear relationships unchanged, and our claim is proved.

2.4 Moebius Transformations

A projectivity can be described by a linear transformation of projective coordinates of the image and preimage line. We arrive at a different formulation if we endow each line with only *one* parameter. So let the preimage point $\underline{\mathbf{x}}$ be given by

$$\underline{\mathbf{x}} = (1-t)\underline{\mathbf{p}}_0 + t\underline{\mathbf{p}}_2$$

and the image $\hat{\underline{\mathbf{x}}}$ by

$$\hat{\underline{\mathbf{x}}} = (1-\hat{t})\hat{\underline{\mathbf{p}}}_0 + \hat{t}\hat{\underline{\mathbf{p}}}_2.$$

If $\underline{\mathbf{p}}_1 = (1-u)\underline{\mathbf{p}}_0 + u\underline{\mathbf{p}}_2$ and $\hat{\underline{\mathbf{p}}}_1 = (1-\hat{u})\hat{\underline{\mathbf{p}}}_0 + \hat{u}\hat{\underline{\mathbf{p}}}_2$, we can define

$$r = \frac{u}{1-u} \quad \text{and} \quad \hat{r} = \frac{\hat{u}}{1-\hat{u}}.$$

The condition of equal cross ratios of image and preimage points now yields

$$r\frac{1-t}{t} = \hat{r}\frac{1-\hat{t}}{\hat{t}},$$

from which we obtain

$$\hat{t} = \frac{\hat{r}t}{(1-t)r + t\hat{r}}. \tag{2.9}$$

This is the desired transformation from t to $\hat{t}$. It is a rational linear transformation, and is often called a *Moebius transformation.*[3] Note that if $r = \hat{r}$, we obtain $\hat{t} = t$.

2.5 Collineations

We have so far studied maps from a line to another line. Going a step further, we may map all of the projective plane $\mathbb{P}^2$ onto itself, using maps that are called *collineations.*

A collineation Φ takes points to points and lines to lines. Moreover, it takes collinear points to collinear points and concurrent lines to concurrent lines. With these definitions, we may already conclude that Φ has a matrix expression when applied to points:

$$\Phi\underline{\mathbf{x}} = M\underline{\mathbf{x}}, \tag{2.10}$$

where M is a nonsingular 3×3 matrix. For if three points $\underline{\mathbf{x}}, \underline{\mathbf{y}}, \underline{\mathbf{z}}$, are collinear, we have

$$\begin{aligned} \det[\underline{\mathbf{x}}, \underline{\mathbf{y}}, \underline{\mathbf{z}}] &= 0, \\ \det[\Phi\underline{\mathbf{x}}, \Phi\underline{\mathbf{y}}, \Phi\underline{\mathbf{z}}] &= \det[M\underline{\mathbf{x}}, M\underline{\mathbf{y}}, M\underline{\mathbf{z}}] \\ &= \det M \det[\underline{\mathbf{x}}, \underline{\mathbf{y}}, \underline{\mathbf{z}}] \\ &= 0. \end{aligned}$$

We can also apply Φ to lines, but now, using the matrix M from (2.10), it takes the form

$$\Phi\underline{\mathbf{L}} = \underline{\mathbf{L}}M^{-1}. \tag{2.11}$$

This is seen by observing that if $\underline{\mathbf{x}}$ lies on $\underline{\mathbf{L}}$, then $\Phi\underline{\mathbf{x}}$ lies on $\Phi\underline{\mathbf{L}}$. In formulas: if $\underline{\mathbf{L}}\underline{\mathbf{x}} = 0$, then $\Phi\underline{\mathbf{L}}\Phi\underline{\mathbf{x}} = 0$. We would like to find a matrix N such that $\Phi\underline{\mathbf{L}} = \underline{\mathbf{L}}N$. Then

$$\Phi\underline{\mathbf{L}}\Phi\underline{\mathbf{x}} = \underline{\mathbf{L}}NM\underline{\mathbf{x}} = \underline{\mathbf{L}}\underline{\mathbf{x}} = 0. \tag{2.12}$$

It follows that $NM = I$, the identity matrix, and thus $N = M^{-1}$. Of course any nonzero multiple αM or αM^{-1} could also serve as the desired maps. Example 2.1 illustrates.

Collineations leave cross ratios invariant: this follows easily from the linearity properties of matrix multiplication, in the same way we proved (2.8). Thus if a collineation Φ maps a line $\underline{\mathbf{L}}$ to another line $\Phi\underline{\mathbf{L}}$, these two lines are related by a projectivity.

---

[3]The terms rational linear transformation or birational transformation are also used.

Let two points be given by

$$\underline{\mathbf{p}}_0 = \begin{bmatrix} 0 \\ 1 \\ 1 \end{bmatrix} \quad \text{and} \quad \underline{\mathbf{p}}_1 = \begin{bmatrix} 1 \\ 0 \\ 1 \end{bmatrix}.$$

Let a collineation Φ be defined by

$$M = \begin{bmatrix} 1 & 0 & 1 \\ 0 & 1 & 0 \\ 0 & 0 & 1 \end{bmatrix},$$

such that

$$\Phi\underline{\mathbf{p}}_0 = \begin{bmatrix} 1 \\ 1 \\ 1 \end{bmatrix} \quad \text{and} \quad \Phi\underline{\mathbf{p}}_1 = \begin{bmatrix} 2 \\ 0 \\ 1 \end{bmatrix}.$$

The line through $\underline{\mathbf{p}}_0$ and $\underline{\mathbf{p}}_1$ is $\underline{\mathbf{L}} = [1, 1, -1]$, and the line through $\Phi\underline{\mathbf{p}}_0, \Phi\underline{\mathbf{p}}_1$ is $\Phi\underline{\mathbf{L}} = [1, 1, -2]$. Since

$$M^{-1} = \begin{bmatrix} 1 & 0 & -1 \\ 0 & 1 & 0 \\ 0 & 0 & 1 \end{bmatrix},$$

we have that $\Phi\underline{\mathbf{L}} = \underline{\mathbf{L}}M^{-1}$.

Example 2.1. Collineations of points and lines.

2.6 Constructing Collineations

How many image and preimage point pairs determine a collineation? Since a collineation is determined by a 3×3 matrix, one might guess that three such pairs are sufficient. But the matrix is only determined up to a constant, and the correct answer is four. We give a geometric demonstration first. Then we will show how to find the matrix that is associated with the collineation.

Referring to Figure 2.5, we assume that we are given four points $\underline{\mathbf{a}}, \underline{\mathbf{b}}, \underline{\mathbf{c}}, \underline{\mathbf{d}}$ and their images. For a given point $\underline{\mathbf{x}}$, we would like to find its image $\hat{\underline{\mathbf{x}}}$.

First, we construct $\underline{\mathbf{d}}_a = [\underline{\mathbf{a}} \wedge \underline{\mathbf{d}}] \wedge [\underline{\mathbf{b}} \wedge \underline{\mathbf{c}}]$. We find $\hat{\underline{\mathbf{d}}}_a$ from the analogous construction with the image points. Next, we construct $\underline{\mathbf{x}}_a = [\underline{\mathbf{a}} \wedge \underline{\mathbf{x}}] \wedge [\underline{\mathbf{b}} \wedge \underline{\mathbf{c}}]$. Its image point $\hat{\underline{\mathbf{x}}}_a$ is found from the condition that

$$\mathrm{cr}(\underline{\mathbf{b}}, \underline{\mathbf{d}}_a, \underline{\mathbf{x}}_a, \underline{\mathbf{c}}) = \mathrm{cr}(\hat{\underline{\mathbf{b}}}, \hat{\underline{\mathbf{d}}}_a, \hat{\underline{\mathbf{x}}}_a, \hat{\underline{\mathbf{c}}}),$$

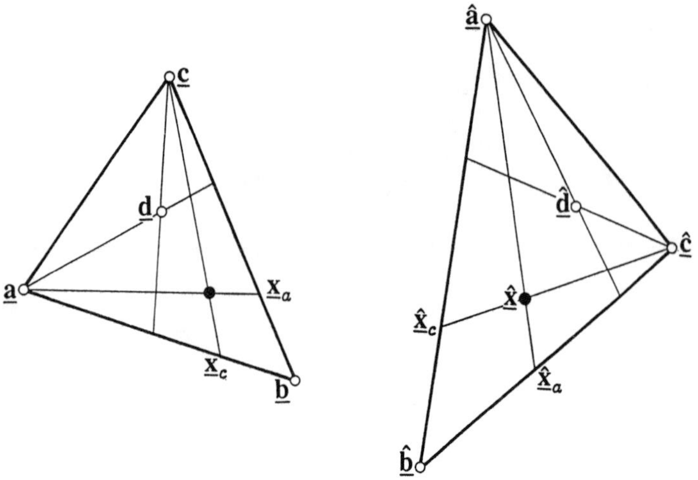

Figure 2.5. Collineations: the image $\hat{\underline{\mathbf{x}}}$ of $\underline{\mathbf{x}}$ may be found by two cross ratio constructions.

by the use of (2.7). In the same manner, we find points $\underline{\mathbf{d}}_c, \hat{\underline{\mathbf{d}}}_c$ and $\underline{\mathbf{x}}_c, \hat{\underline{\mathbf{x}}}_c$. Since $\underline{\mathbf{x}} = [\underline{\mathbf{a}} \wedge \underline{\mathbf{x}}_a] \wedge [\underline{\mathbf{c}} \wedge \underline{\mathbf{x}}_c]$, and since Φ preserves this relationship, we find $\underline{\mathbf{x}}$'s image as

$$\hat{\underline{\mathbf{x}}} = [\hat{\underline{\mathbf{a}}} \wedge \hat{\underline{\mathbf{x}}}_a] \wedge [\hat{\underline{\mathbf{c}}} \wedge \hat{\underline{\mathbf{x}}}_c]. \tag{2.13}$$

Note that $\underline{\mathbf{d}}$ does not have to be "inside" the triangle $\underline{\mathbf{a}}, \underline{\mathbf{b}}, \underline{\mathbf{c}}$ – it just should not be on any of its edges.

What is the matrix M that carries four preimage points to four image points? In order to answer this question, recall that we can always[4] write $\underline{\mathbf{d}}$ as $\underline{\mathbf{d}} \hat{=} \underline{\mathbf{a}} + \underline{\mathbf{b}} + \underline{\mathbf{c}}$ by an appropriate scaling of coordinates. Thus

$$\underline{\mathbf{d}} = \alpha\underline{\mathbf{a}} + \beta\underline{\mathbf{b}} + \gamma\underline{\mathbf{c}}, \tag{2.14}$$

where α, β, γ are those appropriate factors. Equation (2.14) constitutes a three by three linear system for α, β, γ which is solvable. We can now define a projective map of the preimage plane to itself by means of the matrix $A = [\alpha\underline{\mathbf{a}}, \beta\underline{\mathbf{b}}, \gamma\underline{\mathbf{c}}]^{-1}$.

Similarly, we define a matrix $\hat{A} = [\hat{\alpha}\hat{\underline{\mathbf{a}}}, \hat{\beta}\hat{\underline{\mathbf{b}}}, \hat{\gamma}\hat{\underline{\mathbf{c}}}]^{-1}$ which maps the image plane to itself. The numbers $\hat{\alpha}, \hat{\beta}, \hat{\gamma}$ are the solution of the system

$$\hat{\underline{\mathbf{d}}} = \hat{\alpha}\hat{\underline{\mathbf{a}}} + \hat{\beta}\hat{\underline{\mathbf{b}}} + \hat{\gamma}\hat{\underline{\mathbf{c}}}. \tag{2.15}$$

[4]We assume that we are not dealing with degenerate situations here!

Now the preimage point $\underline{\mathbf{x}}$ has been mapped to the same coordinate triple as its image $\hat{\underline{\mathbf{x}}}$:

$$A\underline{\mathbf{x}} \hat{=} \hat{A}\hat{\underline{\mathbf{x}}}.$$

Consequently, since $\hat{\underline{\mathbf{x}}} = M\underline{\mathbf{x}}$ holds for all $\underline{\mathbf{x}}$, we have

$$A = \hat{A}M$$

and thus

$$M = \hat{A}^{-1}A = [\hat{\alpha}\hat{\underline{\mathbf{a}}}, \hat{\beta}\hat{\underline{\mathbf{b}}}, \hat{\gamma}\hat{\underline{\mathbf{c}}}][\alpha\underline{\mathbf{a}}, \beta\underline{\mathbf{b}}, \gamma\underline{\mathbf{c}}]^{-1}. \tag{2.16}$$

At first sight, it appears as if M depends on our initial ordering of the image and preimage points. In fact, M is uniquely determined (up to a constant factor), but we omit the algebra necessary for a proof of this statement.

Example 2.2 shows an application of this technique: it maps a trapezoid onto a square. The three-dimensional analogue of this particular map is known as the *viewing transformation* in computer graphics; see [55]. There, it maps a truncated pyramid onto a cube.

We close this section with a special collineation. In Figure 2.5, assume that $\underline{\mathbf{a}}, \underline{\mathbf{b}}, \underline{\mathbf{c}}$ are mapped to themselves, only $\underline{\mathbf{d}}$ is mapped to a different point $\hat{\underline{\mathbf{d}}}$, but such that $\underline{\mathbf{d}}, \hat{\underline{\mathbf{d}}}, \underline{\mathbf{a}}$ are collinear; see Figure 2.6. The corresponding collineation is called *central.*

Defining $\underline{\mathbf{A}} = \underline{\mathbf{b}} \wedge \underline{\mathbf{c}}$, we see that all points of $\underline{\mathbf{A}}$ are mapped to themselves: we say that $\underline{\mathbf{A}}$ is fixed pointwise by the central collineation. Likewise, all lines in the pencil through $\underline{\mathbf{a}}$ are mapped to themselves: we say that $\underline{\mathbf{a}}$ is fixed linewise. Thus a central collineation is defined by its *center* $\underline{\mathbf{a}}$ and by its *axis* $\underline{\mathbf{A}}$.

2.7 Affine Maps

In the extended affine plane, we may observe the effect that a projective map has on the line at infinity $\mathbf{L}_\infty$. A projective map that maps $\mathbf{L}_\infty$ to itself is called an *affine map.* Note that not every point on $\mathbf{L}_\infty$ needs to be mapped to itself; all we require is that its image be *somewhere* on $\mathbf{L}_\infty$.

The line at infinity corresponds to the projective points $[x, y, 0]^{\mathrm{T}}$, which correspond to affine vectors. Since that line is mapped to itself, vectors are mapped to vectors, and so affine points $[x, y, 1]^{\mathrm{T}}$ must also be mapped to affine points:

$$\Phi : \begin{bmatrix} \star \\ \star \\ 1 \end{bmatrix} \to \begin{bmatrix} \star \\ \star \\ 1 \end{bmatrix} \quad \text{and} \quad \Phi : \begin{bmatrix} \star \\ \star \\ 0 \end{bmatrix} \to \begin{bmatrix} \star \\ \star \\ 0 \end{bmatrix}, \tag{2.17}$$

where $\star$ denotes arbitrary real numbers.

Let four preimage points $\underline{\mathbf{a}}, \underline{\mathbf{b}}, \underline{\mathbf{c}}, \underline{\mathbf{d}}$ be given by

$$\begin{bmatrix} 0 \\ 1 \\ 1 \end{bmatrix}, \begin{bmatrix} 0 \\ -1 \\ 1 \end{bmatrix}, \begin{bmatrix} 1 \\ 2 \\ 1 \end{bmatrix}, \begin{bmatrix} 1 \\ -2 \\ 1 \end{bmatrix},$$

and their respective images by

$$\begin{bmatrix} 0 \\ 2 \\ 1 \end{bmatrix}, \begin{bmatrix} 0 \\ -2 \\ 1 \end{bmatrix}, \begin{bmatrix} 1 \\ 2 \\ 1 \end{bmatrix}, \begin{bmatrix} 1 \\ -2 \\ 1 \end{bmatrix}.$$

The condition $\alpha\underline{\mathbf{a}}+\beta\underline{\mathbf{b}}+\gamma\underline{\mathbf{c}} = \underline{\mathbf{d}}$ leads to a linear system with the solution $\alpha = -2, \beta = 2, \gamma = 1$. The inverse of the matrix $[\alpha\underline{\mathbf{a}}, \beta\underline{\mathbf{b}}, \gamma\underline{\mathbf{c}}]$ is

$$\begin{bmatrix} 0 & 0 & 1 \\ -2 & -2 & 2 \\ -2 & 2 & 1 \end{bmatrix}^{-1} = \frac{1}{4}\begin{bmatrix} 3 & -1 & -1 \\ 1 & -1 & 1 \\ 4 & 0 & 0 \end{bmatrix}.$$

Similarly, we find $\hat{\alpha} = -1, \hat{\beta} = 1, \hat{\gamma} = 1$. Using (2.16), we find

$$M = \frac{1}{4}\begin{bmatrix} 4 & 0 & 0 \\ 0 & 4 & 0 \\ 2 & 0 & 2 \end{bmatrix}.$$

It is easy to check that indeed $M\underline{\mathbf{a}} \hat{=} \underline{\hat{\mathbf{a}}}$ etc. Also, note that the point $[-1, 0, 1]^{\mathrm{T}}$ is mapped to the point $[-1, 0, 0]^{\mathrm{T}}$ at infinity.

Example 2.2. Constructing collineations.

These restrictions on Φ imply restrictions on the matrix M: we find that M's last row must be of the form

$$[m_{31}, m_{32}, m_{33}] = [0, 0, 1]$$

or any nonzero multiple of it. We can thus rewrite M as

$$M = \begin{bmatrix} a_{11} & a_{12} & v_1 \\ a_{21} & a_{22} & v_2 \\ 0 & 0 & 1 \end{bmatrix}.$$

With

$$\underline{\mathbf{x}} = \begin{bmatrix} x \\ y \\ 1 \end{bmatrix} = \begin{bmatrix} \mathbf{x} \\ 1 \end{bmatrix},$$

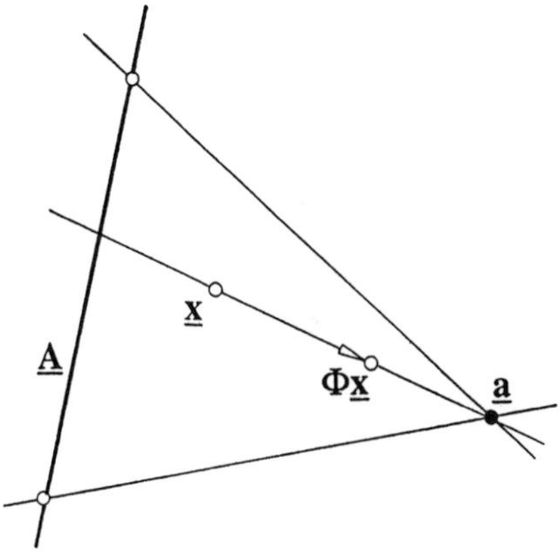

Figure 2.6. Central collineations: these are collineations that are defined by an axis $\underline{\mathbf{A}}$ and a center $\underline{\mathbf{a}}$.

we now have

$$M\underline{\mathbf{x}} = \begin{bmatrix} a_{11} & a_{12} & v_1 \\ a_{21} & a_{22} & v_2 \\ 0 & 0 & 1 \end{bmatrix} \begin{bmatrix} x \\ y \\ 1 \end{bmatrix} = \begin{bmatrix} A\mathbf{x} + \mathbf{v} \\ 1 \end{bmatrix},$$

with $\mathbf{v} = [v_1, v_2]^{\mathrm{T}}$ and $A = \{a_{ij}\}$. Thus an affine map (of the affine plane to itself) is given by the *point map*

$$\Phi\mathbf{x} = A\mathbf{x} + \mathbf{v}; \quad \mathbf{x} \in \mathbb{E}^2, \ \mathbf{v} \in \mathbb{R}^2, \tag{2.18}$$

and the *vector map*

$$\Phi\mathbf{v} = A\mathbf{v}; \quad \mathbf{v} \in \mathbb{R}^2, \tag{2.19}$$

both being obtained in the same way. The vector $\mathbf{v}$ in (2.18) is called the *translational part* of Φ.

A collineation may be defined by the preimages and images of four given lines. Since affine maps always map one of these lines, namely $\mathbf{L}_\infty$, to itself, an affine map is only determined by the preimages and images of three given lines. Three lines intersect in three points, and so we may also think of an affine map as being determined by three preimage/image point pairs.

We pointed out that the fundamental invariant of projective maps is the cross ratio. Affine maps, being special projective maps, leave cross ratios constant, too. But they leave another quantity invariant as well, namely the *ratio* between three collinear points. It is defined as follows: if $\mathbf{b} = (1-\alpha)\mathbf{a} + \alpha\mathbf{c}$, then

$$\text{ratio}(\mathbf{a}, \mathbf{b}, \mathbf{c}) = \frac{\alpha}{1-\alpha}. \tag{2.20}$$

As might be expected, ratios are a special case of cross ratios. If $\mathbf{d}$ is a point at infinity, then

$$\text{ratio}(\mathbf{a}, \mathbf{b}, \mathbf{c}) = \mathbf{cr}(\mathbf{a}, \mathbf{b}, \mathbf{c}, \mathbf{d}).$$

We will encounter ratios in the context of rational Bézier curves; see Section 7.1.

2.8 Problems

1. Show that two perspectivities are not sufficient to describe a projectivity of a line onto itself.

2. Let $\underline{\mathbf{L}}$ be a line with three points $\underline{\mathbf{a}}, \underline{\mathbf{b}}, \underline{\mathbf{c}}$ on it. Let $\underline{\mathbf{x}}$ trace out all of $\underline{\mathbf{L}}$. We define four cross ratios at each point on $\underline{\mathbf{L}}$:

$$\begin{aligned} &\text{cr}(\underline{\mathbf{x}}, \underline{\mathbf{a}}, \underline{\mathbf{b}}, \underline{\mathbf{c}}),\\ &\text{cr}(\underline{\mathbf{a}}, \underline{\mathbf{x}}, \underline{\mathbf{b}}, \underline{\mathbf{c}}),\\ &\text{cr}(\underline{\mathbf{a}}, \underline{\mathbf{b}}, \underline{\mathbf{x}}, \underline{\mathbf{c}}),\\ &\text{cr}(\underline{\mathbf{a}}, \underline{\mathbf{b}}, \underline{\mathbf{c}}, \underline{\mathbf{x}}). \end{aligned}$$

   Plot four graphs: make $\underline{\mathbf{L}}$ the $x-$axis of a standard Euclidean coordinate system and plot the cross ratio values in the y direction.

3. What are the affine analogues of the constructions described by (2.7) and (2.13)?

4. What is the dual of the complete quadrilateral construction from Section 2.2?

5. Draw the dual of Figure 2.5.

6. If affine maps of points are defined by (2.18), how are affine maps of lines defined?

7. What is the projective preimage of the affine zero vector?

3

Conics

Conics are the ancestors of all NURBS. By studying conics, we can observe most of the basic NURBS principles.

The classical definition generates conics as the intersections of a cone with a plane. One then obtains the familiar ellipses, hyperbolas, and parabolas. In projective geometry, this distinction is nonexistent, as is demonstrated in Figure 3.1. There, the tip of the (double) cone is at the origin of three-space, and the cone is intersected with a 3-D plane to yield a conic section in three-space. This conic section is projected into the plane $z = 1$, yielding yet another conic section. But we might have intersected our cone with any other plane, yielding a different conic section in three-space, and the projection would again be the same conic in the $z = 1$ plane! This is the reason — informally stated — why in projective geometry, we cannot distinguish between different types of conics.

3.1 Line Conics

We will not use the above approach to conics, but rather follow an approach that is attributed to J. Steiner:

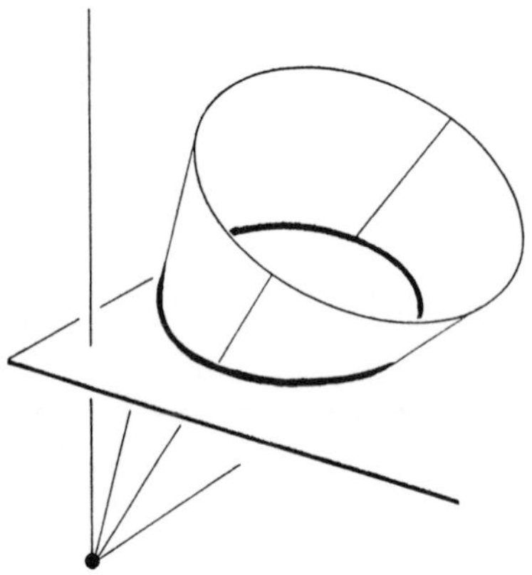

Figure 3.1. Conic sections: the classical definition intersects a cone with a plane.

Let Φ be a projectivity that maps the line $\underline{\mathbf{L}}$ onto another line $\underline{\mathbf{M}}$. We generate a line $\underline{\mathbf{X}}$ by taking a point $\underline{\mathbf{x}}$ on $\underline{\mathbf{L}}$, constructing its image $\hat{\underline{\mathbf{x}}} = \Phi\underline{\mathbf{x}}$, and by connecting the two points:

$$\underline{\mathbf{X}} = \underline{\mathbf{x}} \wedge \hat{\underline{\mathbf{x}}}. \tag{3.1}$$

The collection of all such lines $\underline{\mathbf{X}}$ is called a *conic* Γ. The conic is called *singular* if either $\underline{\mathbf{L}} = \underline{\mathbf{M}}$ or if Φ is a perspectivity. The two lines $\underline{\mathbf{L}}$ and $\underline{\mathbf{M}}$ are called *generating lines*.

If Φ is a perspectivity, the conic degenerates to a point, namely the center of the perspectivity. If the two generating lines coincide, the conic degenerates to that line. If a line $\underline{\mathbf{X}}$ is one of the lines that define Γ, we say that $\underline{\mathbf{X}}$ is on Γ; we also say that $\underline{\mathbf{X}}$ is a *tangent* of Γ, a notion that will be justified below.

Since Φ is a projectivity, it is a bijective map. So we might as well have used Φ^{-1} in our definition. Figure 3.2 illustrates.

A dual definition also exists. It defines a conic as the locus of points, namely all intersections of corresponding lines in two projectively related pencils; see Figure 3.2. Such conics are called *point conics*, whereas our definition would generate *line conics*. Both definitions yield the same class of curves, as we shall see later.

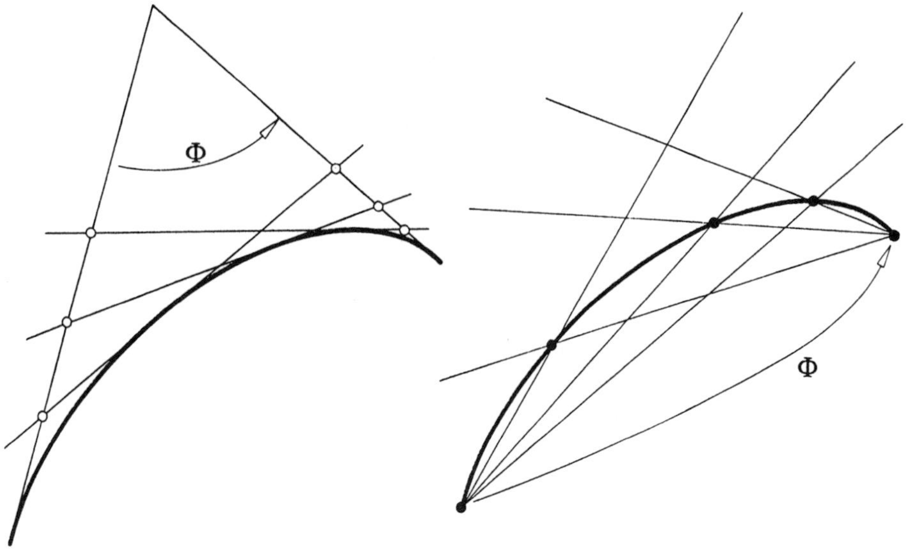

Figure 3.2. Conics: left: line conic; right: point conic.

Returning to line conics, it appears that a conic Γ depends crucially on the generating lines $\underline{\mathbf{L}}$ and $\underline{\mathbf{M}}$. This is not so: any other two lines on it can serve equally well as generating lines.[1] In order to see this, let $\underline{\mathbf{A}}$ be a line on the conic with generating lines $\underline{\mathbf{L}}$ and $\underline{\mathbf{M}}$. Our claim is now that we can also make $\underline{\mathbf{A}}$ and $\underline{\mathbf{M}}$ generating lines of Γ. Referring to Figure 3.3, we need projectively related ranges on $\underline{\mathbf{A}}$ and $\underline{\mathbf{M}}$ such that the connection of corresponding points are lines on Γ. A projectivity from one range to another is determined by three image and preimage pairs. If we can produce three such pairs, we have proved our claim. We can choose

$$
\begin{array}{ccccccc}
\underline{\mathbf{a}} & \text{on} & \underline{\mathbf{A}} & \text{and} & \underline{\mathbf{L}} \wedge \underline{\mathbf{M}} & \text{on} & \underline{\mathbf{M}}, \\
\underline{\mathbf{B}} \wedge \underline{\mathbf{A}} & \text{on} & \underline{\mathbf{A}} & \text{and} & \hat{\underline{\mathbf{b}}} & \text{on} & \underline{\mathbf{M}}, \\
\underline{\mathbf{C}} \wedge \underline{\mathbf{A}} & \text{on} & \underline{\mathbf{A}} & \text{and} & \hat{\underline{\mathbf{c}}} & \text{on} & \underline{\mathbf{M}},
\end{array}
$$

where $\underline{\mathbf{A}} = \underline{\mathbf{a}} \wedge \hat{\underline{\mathbf{a}}}$, etc.

Obviously we can therefore make any two lines on a conic its generating lines. It also follows that the lines $\underline{\mathbf{L}}$ and $\underline{\mathbf{M}}$ themselves are on the conic.

As a consequence, a line conic is defined by five distinct lines; dually, a point conic is defined by five distinct points.

[1]This is a theoretical statement: in a "real" situation, not all generating lines are numerically equivalent!

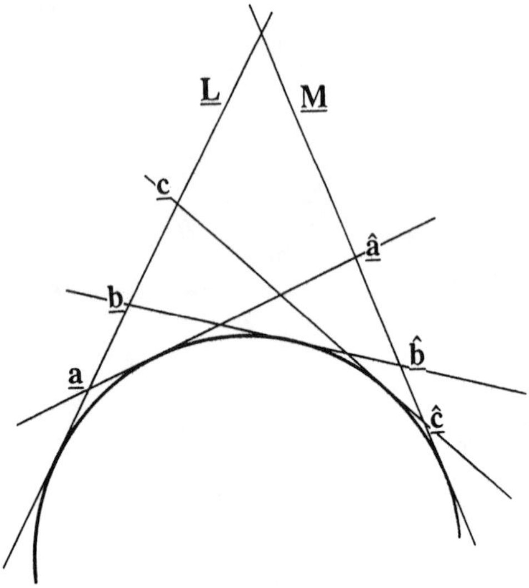

Figure 3.3. Generating lines: any two of the shown lines can be taken as generating lines.

At this point, we do not know much about the shape of a conic — all we know is that a conic is defined by a set of lines or, dually, by a set of points. But we can already make one claim: a nonsingular point conic may not be intersected by any line in more than two points.[2] For if such a line $\underline{\mathbf{K}}$ existed, it would have three pairs of preimage/image lines intersecting on it. But then the projectivity between the two pencils must be a perspectivity, and the conic would be singular.

Dually, we can say that a nonsingular line conic may not have more than two tangents through any point $\underline{\mathbf{p}}$ not on the conic.

3.2 Conic Constructions

It is now time to become more concrete: we know what it takes for a line to *be* on a conic, but how can we *find* one? Recall that a projectivity between two lines is defined by three image/preimage point pairs. Pick three points $\underline{\mathbf{a}}, \underline{\mathbf{b}}, \underline{\mathbf{c}}$ on $\underline{\mathbf{L}}$ and $\hat{\underline{\mathbf{a}}}, \hat{\underline{\mathbf{b}}}, \hat{\underline{\mathbf{c}}}$ on $\underline{\mathbf{M}}$, such that their connecting lines are on Γ.

[2]A point conic is intersected by a line in a point $\underline{\mathbf{p}}$ if $\underline{\mathbf{p}}$ is both on the conic and on the line.

How can we find more lines on Γ?

If $\underline{\mathbf{x}}$ is an arbitrary point on $\underline{\mathbf{L}}$, we have found a line on the conic as soon as we have found its image $\hat{\underline{\mathbf{x}}}$ on $\underline{\mathbf{M}}$ under the projectivity Φ that is defined by our three point pairs. But we already discussed how to find the image of an arbitrary point under a projectivity: this was done in Section 2.3 by means of Pappus' theorem. We would proceed exactly as in Figure 2.4.

We may state this construction slightly differently: given five lines on a conic, find a sixth one. The given lines are $\underline{\mathbf{M}}, \underline{\mathbf{L}}, \underline{\mathbf{a}} \wedge \hat{\underline{\mathbf{a}}}, \underline{\mathbf{b}} \wedge \hat{\underline{\mathbf{b}}}$, and $\underline{\mathbf{c}} \wedge \hat{\underline{\mathbf{c}}}$.

The dual definition of conics as point conics leads to the dual construction; it is shown in Figure 3.4.

There, three concurrent lines $\underline{\mathbf{A}}, \underline{\mathbf{B}}, \underline{\mathbf{C}}$ through $\underline{\mathbf{l}}$ and their projective images $\hat{\underline{\mathbf{A}}}, \hat{\underline{\mathbf{B}}}, \hat{\underline{\mathbf{C}}}$ through $\underline{\mathbf{m}}$ are given. For an arbitrary line $\underline{\mathbf{X}}$ through $\underline{\mathbf{l}}$, we want to find its image $\hat{\underline{\mathbf{X}}}$, so that we can find $\underline{\mathbf{x}} = \underline{\mathbf{X}} \wedge \hat{\underline{\mathbf{X}}}$ on the conic. The dual of Pappus' theorem provides a construction: First, we construct the Pappus point $\underline{\mathbf{p}}$ (see Section 1.8). Then we set

$$\underline{\mathbf{Y}} = [\hat{\underline{\mathbf{A}}} \wedge \underline{\mathbf{X}}] \wedge \underline{\mathbf{p}},$$

and find $\hat{\underline{\mathbf{X}}}$ as

$$\hat{\underline{\mathbf{X}}} = [\underline{\mathbf{A}} \wedge \underline{\mathbf{Y}}] \wedge \underline{\mathbf{m}}.$$

This construction may be viewed as a method to construct a sixth point on a conic once five points on it are given. This is a classical construction, and has been used by generations of draftsmen; see [5], [79], [78].

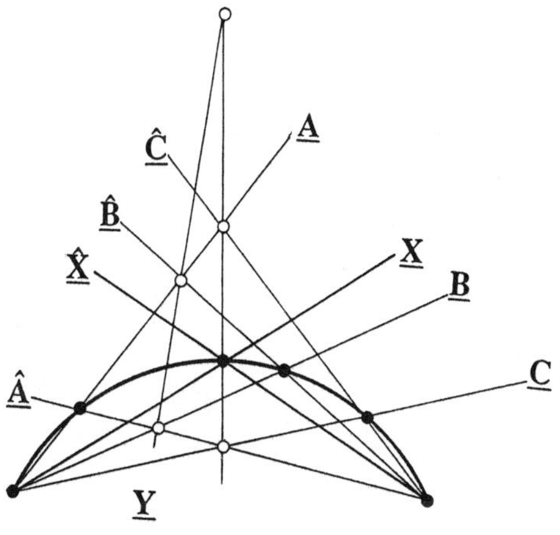

Figure 3.4. Point conic construction: a point on the conic is found by using the dual of Pappus' theorem.

3.3 Points and Lines

What does it take for a point to be on a line conic? This very concept is not even defined yet, so here is a definition: A point $\underline{\mathbf{x}}$ is on a line conic Γ if there is exactly one line $\underline{\mathbf{L}}$ on Γ that contains $\underline{\mathbf{x}}$.

Dually, we may define a line on a point conic; such lines are the conic's *tangents.* A line $\underline{\mathbf{L}}$ is on a point conic Γ (i.e., $\underline{\mathbf{L}}$ is tangent to Γ), if there is exactly one point on Γ that lies on $\underline{\mathbf{L}}$.[3]

If a (line) conic Γ is defined by a projectivity Φ between two generating lines $\underline{\mathbf{L}}$ and $\underline{\mathbf{M}}$, then we know that those two lines are themselves on Γ.[4] We now ask: where does the conic touch $\underline{\mathbf{L}}$ and $\underline{\mathbf{M}}$, i.e., which are the points on the generating lines that are also on the conic? Let $\underline{\mathbf{b}}_1$ be the intersection of $\underline{\mathbf{L}}$ and $\underline{\mathbf{M}}$, as shown in Figure 3.5. Since $\underline{\mathbf{b}}_1$ is on $\underline{\mathbf{L}}$, its image $\Phi\underline{\mathbf{b}}_1$ is on $\underline{\mathbf{M}}$. We call this image point $\underline{\mathbf{b}}_2$. Conversely, $\underline{\mathbf{b}}_1$ is also on $\underline{\mathbf{M}}$, and so its image $\Phi^{-1}\underline{\mathbf{b}}_1$ is on $\underline{\mathbf{L}}$. We define $\underline{\mathbf{b}}_0 = \Phi^{-1}\underline{\mathbf{b}}_1$.

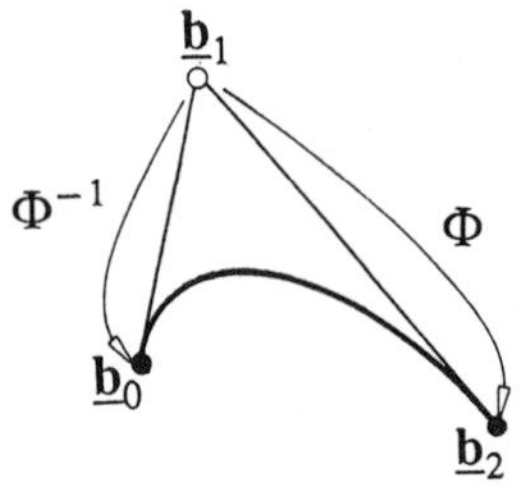

Figure 3.5. Points on a line conic: $\underline{\mathbf{b}}_0$ and $\underline{\mathbf{b}}_2$ lie on Γ.

We now claim that $\underline{\mathbf{b}}_0$ and $\underline{\mathbf{b}}_2$ are the two desired points of contact. Concentrating on $\underline{\mathbf{b}}_0$, we must show that a) $\underline{\mathbf{b}}_0$ is on Γ, and b) there is no other line on Γ that contains $\underline{\mathbf{b}}_0$. Since $\underline{\mathbf{b}}_0$ is the image of $\underline{\mathbf{b}}_1$ under Φ^{-1}, a) is satisfied. If another line on Γ existed that contains $\underline{\mathbf{b}}_0$, a second point – other than $\underline{\mathbf{b}}_1$ – must exist such that its image under Φ^{-1} is $\underline{\mathbf{b}}_0$. This is impossible, since Φ^{-1} is a bijection, and so our claim is verified.

Since we can generate a line conic from any two projectively related ranges, we are now able to find the *point of contact* on any line on the conic. For example, if we were given five lines to a conic, we are now able to find where they touch the conic. With five points on the conic, we may treat it as a point conic: thus we have shown that *every line conic can be expressed as a point conic and vice versa.* The "vice versa" part in this equivalence statement addresses the dual situation, of course. With this knowledge, we may refer to "lines on a conic" as "tangents to a conic."

[3]Note how we avoided any calculus here!

[4]See, for example, Figure 3.3.

In particular, we can state that a conic is defined by

a) two tangents $\underline{\mathbf{T}}_0, \underline{\mathbf{T}}_1$,

b) their points of contact $\underline{\mathbf{b}}_0, \underline{\mathbf{b}}_2$,

c) another tangent, defined by two points $\underline{\mathbf{q}}_0$ and $\underline{\mathbf{q}}_1$ on $\underline{\mathbf{T}}_0$ and $\underline{\mathbf{T}}_1$, respectively.

This tangent from c) is called the *shoulder tangent.* With $\underline{\mathbf{b}}_1$ being the intersection of $\underline{\mathbf{T}}_0$ and $\underline{\mathbf{T}}_1$, all we need to do is establish a projectivity between $\underline{\mathbf{T}}_0$ and $\underline{\mathbf{T}}_1$. It is given by the following preimage/image point pairs:

$$\underline{\mathbf{b}}_0 \to \underline{\mathbf{b}}_1, \quad \underline{\mathbf{q}}_0 \to \underline{\mathbf{q}}_1, \quad \underline{\mathbf{b}}_1 \to \underline{\mathbf{b}}_2.$$

See Figure 3.6 for an illustration and Chapter 4 for further elaboration.

In the context of point conics, we might have replaced the tangent $\underline{\mathbf{q}}_0 \wedge \underline{\mathbf{q}}_1$ by a *shoulder point,* $\underline{\mathbf{q}}$.

We are also in the position to formulate and prove another central theorem: the *four tangent theorem.* Consult Figure 3.7 for the notation used. The theorem goes back to the projective development of conics by J. Steiner.

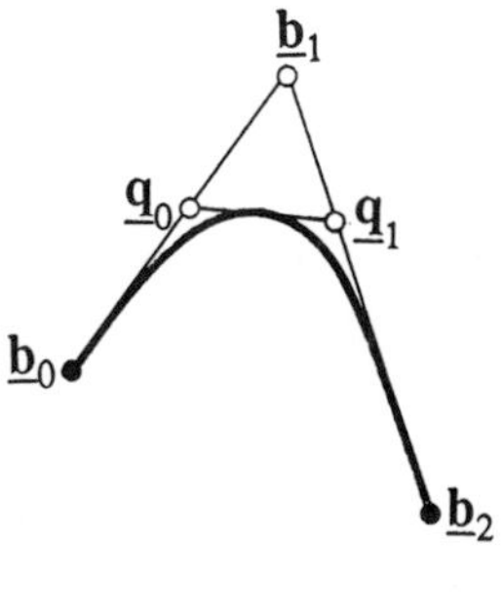

Figure 3.6. Conic constraints: a conic is given by two points, their corresponding tangents, and a third tangent.

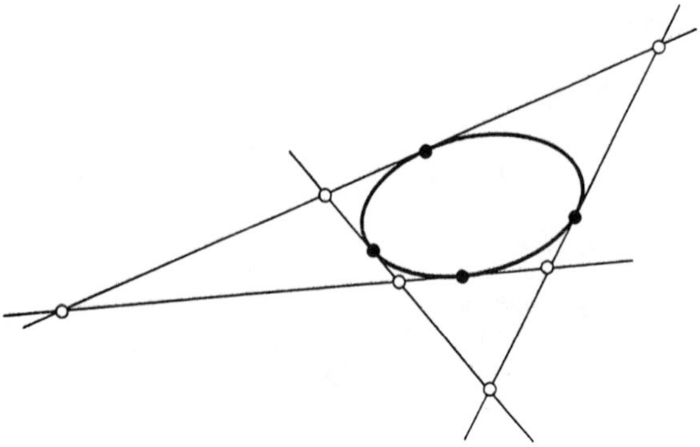

Figure 3.7. The four tangent theorem: four points are shown on each of four tangents to a conic. Their cross ratios are equal.

Let $\underline{\mathbf{L}}_i; i = 1,2,3,4$, be four tangents to a conic Γ, and let $\underline{\mathbf{p}}_{i,i}$ their points of contact with Γ. Set

$$\underline{\mathbf{p}}_{i,j} = \underline{\mathbf{L}}_i \wedge \underline{\mathbf{L}}_j \quad \text{for} \quad i \neq j.$$

Then the following four cross ratios are all equal to a common constant c:

$$\text{cr}(\underline{\mathbf{p}}_{1,j}, \underline{\mathbf{p}}_{2,j}, \underline{\mathbf{p}}_{3,j}, \underline{\mathbf{p}}_{4,j}) = c; \quad j = 1,2,3,4. \tag{3.2}$$

For a proof, we let $\underline{\mathbf{L}}_1$ and $\underline{\mathbf{L}}_4$ be generating lines of the conic. Then the four points on each of them are related by a projectivity. Since any pair of distinct lines $\underline{\mathbf{L}}_i$ and $\underline{\mathbf{L}}_j$ can also be used as generating lines, the theorem is proved.

The dual of the four tangent theorem will be called the *four point theorem* and can be stated like this: Let $\underline{\mathbf{p}}_i; i = 1,2,3,4$, be four points on a (point) conic Γ, and let $\underline{\mathbf{L}}_{i,i}$ be the tangents to Γ at $\underline{\mathbf{p}}_i$. Set

$$\underline{\mathbf{L}}_{i,j} = \underline{\mathbf{p}}_i \wedge \underline{\mathbf{p}}_j \quad \text{for} \quad i \neq j.$$

Then the following four cross ratios are all equal to a common constant c:

$$\text{cr}(\underline{\mathbf{L}}_{1,j}, \underline{\mathbf{L}}_{2,j}, \underline{\mathbf{L}}_{3,j}, \underline{\mathbf{L}}_{4,j}) = c; \quad j = 1,2,3,4. \tag{3.3}$$

This simply says that all four pencils through the $\underline{\mathbf{p}}_i$ are projectively related. For an illustration, see Figure 3.8.

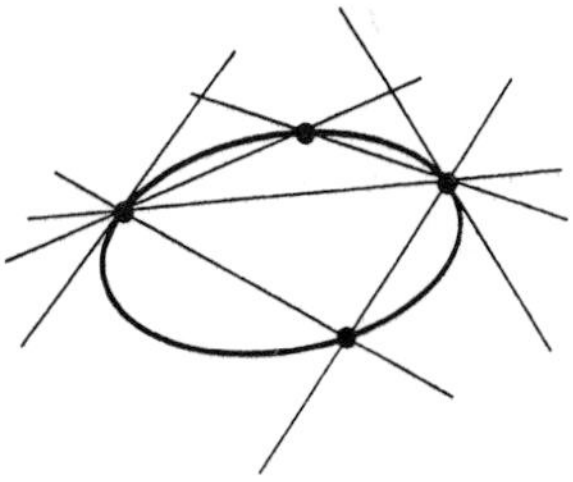

Figure 3.8. The four point theorem: four points are shown as well as the four pencils generated by them. All four pencils are projectively related.

In that figure, let us concentrate on two of the points and the pencils through them, as shown in Figure 3.9. If one pencil intersects a line $\underline{\mathbf{L}}$ and the other intersects a line $\underline{\mathbf{M}}$, both lines are related by a projectivity: if $\underline{\mathbf{x}}$ is a point on $\underline{\mathbf{L}}$, its image $\hat{\underline{\mathbf{x}}}$ is on $\underline{\mathbf{M}}$. We find $\hat{\underline{\mathbf{x}}}$ as follows: connect $\underline{\mathbf{x}}$ with $\underline{\mathbf{q}}$ and record the intersection $\underline{\mathbf{y}}$ of this line with Γ. Join $\underline{\mathbf{y}}$ with $\underline{\mathbf{p}}$, and $\hat{\underline{\mathbf{x}}}$ is the intersection of $\underline{\mathbf{p}} \wedge \underline{\mathbf{y}}$ with $\underline{\mathbf{M}}$.

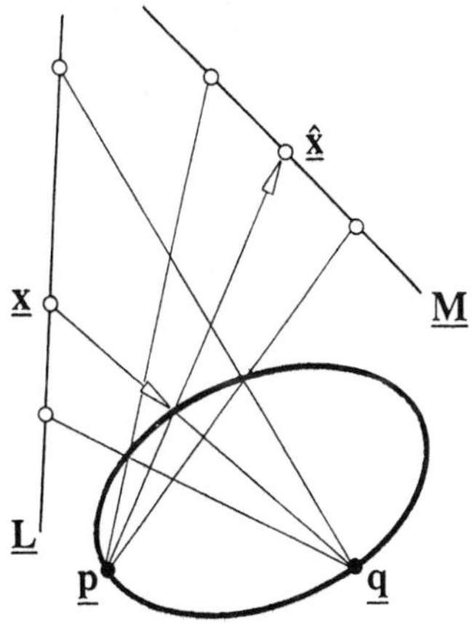

Figure 3.9. Conics and projectivities: the lines $\underline{\mathbf{L}}$ and $\underline{\mathbf{M}}$ are related by a conic-induced projectivity.

3.4 Pascal's Theorem

Pascal's theorem is the first significant (*very* significant) advance that the theory of conics has seen since it became stagnant in the Middle Ages. B.

Pascal, who never received any formal training in mathematics, discovered it at the age of sixteen, in 1641. Pascal's theorem may be stated as follows:

Let $\underline{\mathbf{p}}_1, \underline{\mathbf{p}}_2, \underline{\mathbf{p}}_3$ and $\underline{\mathbf{q}}_1, \underline{\mathbf{q}}_2, \underline{\mathbf{q}}_3$ be any six points on a conic, as shown in Figure 3.10. Then the three points

$$\begin{aligned}
\underline{\mathbf{c}}_1 &= (\underline{\mathbf{p}}_2 \wedge \underline{\mathbf{q}}_3) \wedge (\underline{\mathbf{p}}_3 \wedge \underline{\mathbf{q}}_2),\\
\underline{\mathbf{c}}_2 &= (\underline{\mathbf{p}}_1 \wedge \underline{\mathbf{q}}_3) \wedge (\underline{\mathbf{p}}_3 \wedge \underline{\mathbf{q}}_1),\\
\underline{\mathbf{c}}_3 &= (\underline{\mathbf{p}}_1 \wedge \underline{\mathbf{q}}_2) \wedge (\underline{\mathbf{p}}_2 \wedge \underline{\mathbf{q}}_1)
\end{aligned}$$

are collinear. The line on which they exist is called the *Pascal line.*

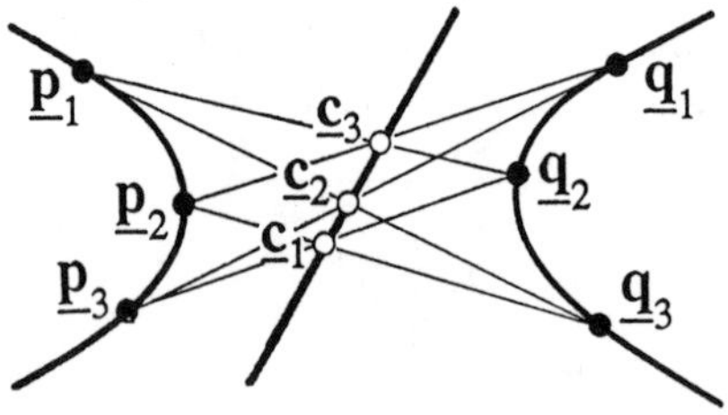

Figure 3.10. Pascal's theorem: the three hollow points are collinear.

For a proof, we turn to Figure 3.11. First, we observe that we can relate the pencils through $\underline{\mathbf{q}}_2$ and $\underline{\mathbf{q}}_3$ by a projectivity; see the four point theorem

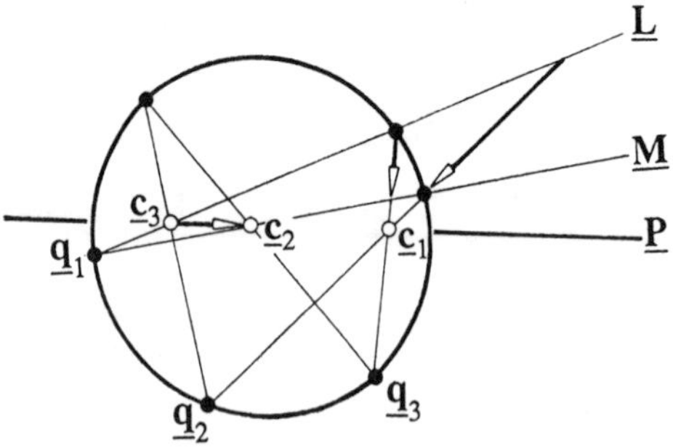

Figure 3.11. Pascal's theorem: the proof proceeds by showing that $\underline{\mathbf{L}}$ and $\underline{\mathbf{M}}$ are related by a perspectivity.

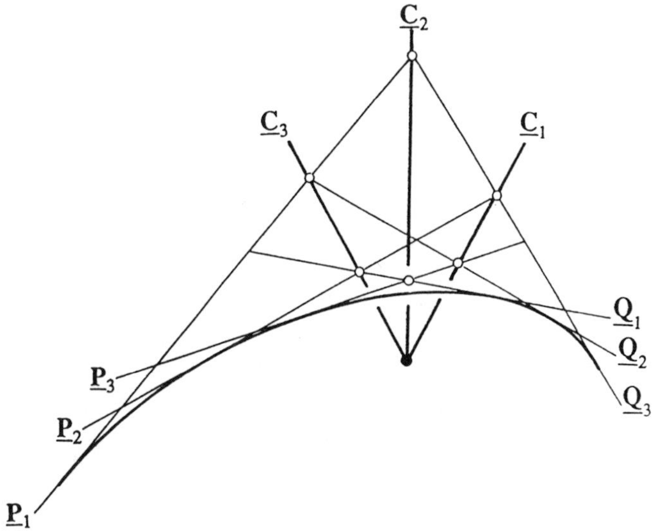

Figure 3.12. Brianchon's theorem: the three lines $\underline{\mathbf{C}}_i$ are concurrent.

above. This in turn generates a projectivity Φ between the lines $\underline{\mathbf{M}}$ and $\underline{\mathbf{L}}$. Since $\Phi\underline{\mathbf{q}}_1 = \underline{\mathbf{q}}_1$, we know that Φ is in fact a perspectivity, see Section 2.1.

Now let us determine $\Phi\underline{\mathbf{c}}_3$. To do this, we connect $\underline{\mathbf{c}}_3$ with $\underline{\mathbf{q}}_2$ and intersect the connecting line with Γ, yielding $\underline{\mathbf{p}}_1$. Then we join $\underline{\mathbf{p}}_1$ with $\underline{\mathbf{q}}_3$ and intersect with $\underline{\mathbf{M}}$: the image of $\underline{\mathbf{c}}_3$ is $\underline{\mathbf{c}}_2$. Thus the center of Φ must lie on $\underline{\mathbf{c}}_1 \wedge \underline{\mathbf{c}}_2$. We also know that $\Phi\underline{\mathbf{p}}_2 \wedge \underline{\mathbf{q}}_3$ must pass through Φ's center, which is therefore $\underline{\mathbf{c}}_1$. Hence the three $\underline{\mathbf{c}}_i$ are collinear. Since $\underline{\mathbf{c}}_1$ is also on the line $\underline{\mathbf{q}}_2 \wedge \underline{\mathbf{p}}_3$, Pascal's theorem is proved.

Pascal's original proof proceeded in a different manner: he proved his theorem for a circle. Any other conic is a projective image of the circle, and the involved constructions are projectively invariant. Thus a proof for the circle is sufficient to prove the general case!

If the six points on the conic are not aligned as in our figure, Pascal's theorem still holds: as we relabel the points, there will be different Pascal lines.

If we dualize Pascal's theorem (see Section 1.8), we obtain *Brianchon's theorem*:

Let $\underline{\mathbf{P}}_1, \underline{\mathbf{P}}_2, \underline{\mathbf{P}}_3$ and $\underline{\mathbf{Q}}_1, \underline{\mathbf{Q}}_2, \underline{\mathbf{Q}}_3$ be any six tangents on a conic, as shown in Figure 3.12. Then the three lines

$$\begin{aligned}
\underline{\mathbf{C}}_1 &= (\underline{\mathbf{P}}_2 \wedge \underline{\mathbf{Q}}_3) \wedge (\underline{\mathbf{P}}_3 \wedge \underline{\mathbf{Q}}_2),\\
\underline{\mathbf{C}}_2 &= (\underline{\mathbf{P}}_1 \wedge \underline{\mathbf{Q}}_3) \wedge (\underline{\mathbf{P}}_3 \wedge \underline{\mathbf{Q}}_1),\\
\underline{\mathbf{C}}_3 &= (\underline{\mathbf{P}}_1 \wedge \underline{\mathbf{Q}}_2) \wedge (\underline{\mathbf{P}}_2 \wedge \underline{\mathbf{Q}}_1)
\end{aligned}$$

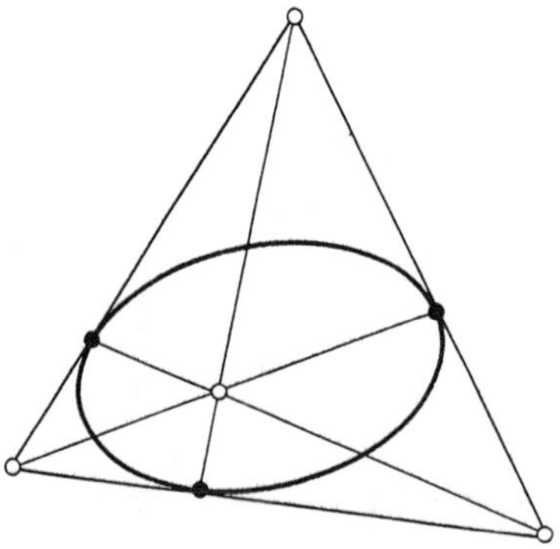

Figure 3.13. Brianchon's theorem: in this special case, a triangle is circumscribed around a conic. The three "medians" are concurrent.

are concurrent. The point in which they meet is called the *Brianchon point.*

We do not have to prove this theorem due to the principle of duality. When Brianchon proved his theorem — in 1806 — this principle had not yet been discovered.

A special case of Brianchon's theorem is obtained when we let the six points approach each other, resulting in three points together with their tangents. We then have: if a triangle is circumscribed around a conic, the lines through points of contact and opposite triangle vertices are concurrent.[5] See Figure 3.13 for an illustration.

Pascal's theorem makes an assertion about six points known to be on a conic. If we retrace the proof of Pascal's theorem "backwards," we see that the following is also true: if six points $\underline{\mathbf{p}}_i, \underline{\mathbf{q}}_i$, $i = 1, 2, 3$, determine three collinear points $\underline{\mathbf{c}}_i = \underline{\mathbf{p}}_j \wedge \underline{\mathbf{p}}_k$, with i, j, k all different, then the six points lie on a conic.

Finally, a special case of Pascal's theorem: referring to Figure 3.14, we let two point pairs coincide: we set $\underline{\mathbf{p}}_1 = \underline{\mathbf{q}}_3$ and $\underline{\mathbf{p}}_3 = \underline{\mathbf{q}}_1$. Then the lines $\underline{\mathbf{p}}_1 \wedge \underline{\mathbf{q}}_3$ and $\underline{\mathbf{p}}_3 \wedge \underline{\mathbf{q}}_1$ become tangents. The Pascal line now goes through the Pappus point $\underline{\mathbf{p}}$ of the two pencils through $\underline{\mathbf{p}}_1 = \underline{\mathbf{q}}_3$ and $\underline{\mathbf{p}}_3 = \underline{\mathbf{q}}_1$.

This leads to one of the classical conic constructions: given three points and two tangents to a conic, find more points on it. Again referring to Figure 3.14, assume the tangents $\underline{\mathbf{T}}_0$ and $\underline{\mathbf{T}}_1$ in $\underline{\mathbf{p}}_1 = \underline{\mathbf{q}}_3$ and $\underline{\mathbf{p}}_3 = \underline{\mathbf{q}}_1$ are given, plus the point $\underline{\mathbf{p}}_2$. We can now fix an arbitrary Pascal line $\underline{\mathbf{L}}$ and then construct the point $\underline{\mathbf{q}}_2$ defined by it. As we let the Pascal line vary,

[5] A simple proof for this special case is as follows: the statement holds for an equilateral triangle circumscribed around a circle. All other cases follow by projectively mapping this configuration.

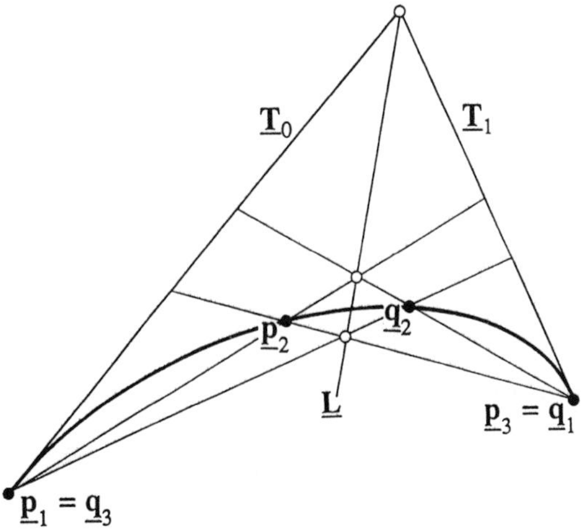

Figure 3.14. Pascal's theorem: a special case.

we generate more points on the conic. Since we can write

$$\underline{\mathbf{L}} = (1-t)\underline{\mathbf{T}}_0 + t\underline{\mathbf{T}}_1,$$

we have found a *parametrization* of our conic: for every value of t, we can find a point $\underline{\mathbf{q}}_2(t)$ on the conic. In the next chapter, we will see a more elegant parametric form for conics.

3.5 Affine Conics

When we consider the extended affine plane, we know that affine maps leave the line $\mathbf{L}_\infty$ invariant, according to the development in Section 2.7. A conic may have zero, one, or two intersections with $\mathbf{L}_\infty$, and this number would be unchanged by an affine map. This leads to the following classification of conics in the extended affine plane: *a conic is called an ellipse, a parabola, or a hyperbola, if it has zero, one, or two intersections with* $\mathbf{L}_\infty$*, respectively.* Since $\mathbf{L}_\infty$ is the image of some line $\underline{\mathbf{H}}$ under a projective map from $\mathbb{P}^2$, we arrive at the Fig. 3.15.

Before we look at each affine conic type, a definition: a curve is called *convex* if, for any tangent to it, the whole curve lies on one side of this tangent. Note that this definition only makes sense in an affine context! See Section 1.10 for an explanation.

The **ellipse**, having no points in common with the line at infinity, is a finite curve. Since it cannot be intersected by any line more than twice, it is a *convex* curve.

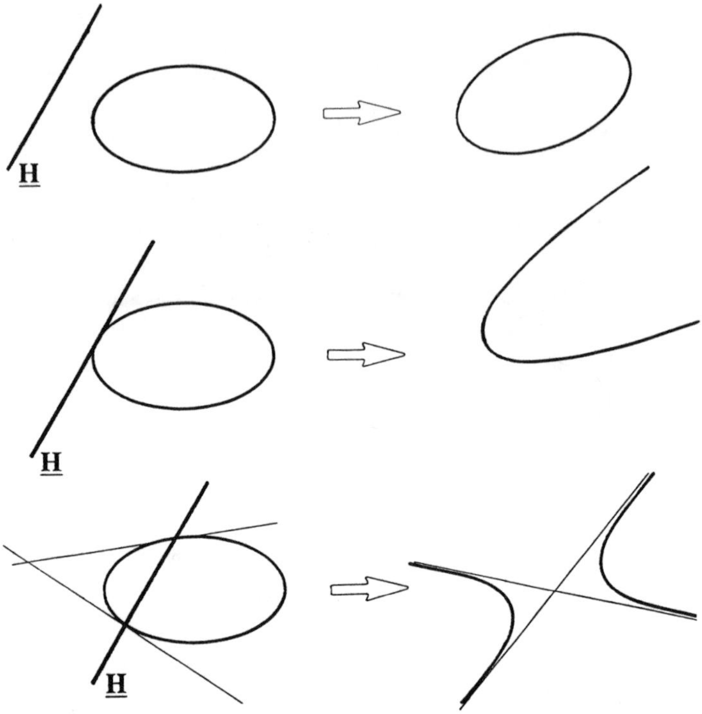

Figure 3.15. Affine conics: from top to bottom: ellipse, parabola, hyperbola. In each case, $\underline{\mathbf{H}}$ is the preimage of the line at infinity.

Four points are sufficient to determine an ellipse. The points must be vertices of a convex quadrilateral. In order to produce a numerically reliable ellipse, the four points should not be "too close" together.

The **parabola** has exactly one point in common with $\mathbf{L}_\infty$, hence $\mathbf{L}_\infty$ is a tangent of the parabola. It is also a convex curve, although not a finite one.

Since we know one tangent of the parabola, four more points $\mathbf{p}_i$ will determine it completely. But these points cannot be located arbitrarily since a parabola is a convex curve. The points must be vertices of a convex quadrilateral; see also [76].

The **hyperbola** intersects $\mathbf{L}_\infty$ twice. Let $\underline{\mathbf{x}}$ and $\underline{\mathbf{y}}$ be the intersections with the hyperbola's preimage with $\underline{\mathbf{H}}$. The tangents at $\underline{\mathbf{x}}$ and $\underline{\mathbf{y}}$ are mapped to the *asymptotes* of the hyperbola. A hyperbola is not a convex curve.

A hyperbola is determined by its two asymptotes and one more point.

3.6 Problems

1. Show that two conics cannot intersect in more than four points.
2. Specialize the situation from Figure 3.14 even more, namely by considering three points on a conic and their respective tangents.
3. Develop the construction of a point conic using a given shoulder point.
4. What are the models of line and point conics in the spherical or circular models of the projective plane? If you have access to the appropriate kind of graphics equipment, illustrate!

4

Conics in Parametric Form

So far, we have covered the basic theoretical aspects of conics in the projective plane. Our definition was based on the concept of projectivities between lines or pencils – it is now time to move to a more tangible description. This will be the parametric form, which will allow us to trace curves, and also to calculate derivatives. We will spend some time on parametric conics, as they are important by themselves, but also since many of their properties will carry over very naturally to rational Bézier curves or to NURBS.

4.1 Parametric Curves in $\mathbb{P}^2$

Before we start to treat conics in parametric form, we present a few facts about parametric curves in projective space; some phenomena arise here that are quite different from the more familiar affine treatment of curves.

A parametric curve $\underline{\mathbf{x}}(t)$ is a map from the projective line — endowed with a parameter t — into $\mathbb{P}^2$. Let us consider the derivative of $\underline{\mathbf{x}}(t)$. We assume that each of the components of $\underline{\mathbf{x}}$ is differentiable with respect to t. Let Δt be a nonzero number. Then the three collinear points $\underline{\mathbf{x}}(t), \underline{\mathbf{x}}(t + \Delta t), \underline{\mathbf{x}}(t + \Delta t) - \underline{\mathbf{x}}(t)$ are on a *secant* of the curve, as shown in Figure 4.1.

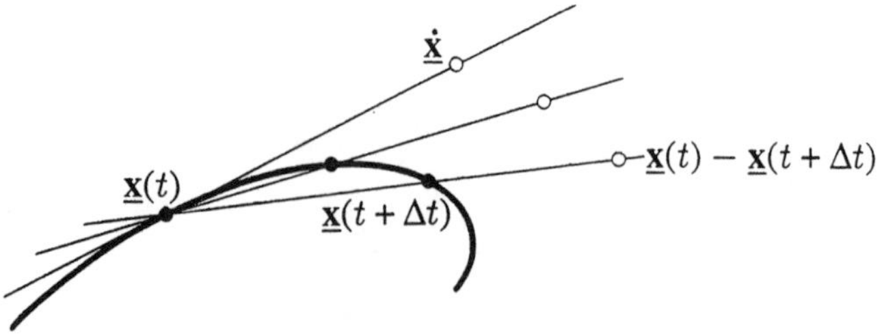

Figure 4.1. Projective derivatives: as $\Delta t \to 0$, the secants approach the tangent at $\underline{\mathbf{x}}(t)$.

Note that

$$\underline{\mathbf{x}}(t+\Delta t) - \underline{\mathbf{x}}(t) \mathrel{\hat{=}} \frac{\underline{\mathbf{x}}(t+\Delta t) - \underline{\mathbf{x}}(t)}{\Delta t}.$$

We now take the limit $\Delta t \to 0$:

$$\dot{\underline{\mathbf{x}}} := \frac{\mathrm{d}\underline{\mathbf{x}}(t)}{\mathrm{d}t} = \lim_{\Delta t \to 0} \frac{\underline{\mathbf{x}}(t+\Delta t) - \underline{\mathbf{x}}(t)}{\Delta t}.$$

In this limit process, the secant approaches the tangent $\underline{\mathbf{T}}(t)$ and the point $\big(\underline{\mathbf{x}}(t+\Delta t) - \underline{\mathbf{x}}(t)\big)/\Delta t$ approaches the *derivative* $\dot{\underline{\mathbf{x}}}(t)$.

We are used to thinking of derivatives as tangent *vectors*, but in the projective plane, there are no vectors, and hence the derivative is yet another point. This point is located on the tangent $\underline{\mathbf{T}}(t)$, which is thus given by

$$\underline{\mathbf{T}} = \underline{\mathbf{x}} \wedge \dot{\underline{\mathbf{x}}}. \tag{4.1}$$

Similarly, the second derivative $\ddot{\underline{\mathbf{x}}}$ is yet another point in $\mathbb{P}^2$. If, for some t, the three points $\underline{\mathbf{x}}, \dot{\underline{\mathbf{x}}}, \ddot{\underline{\mathbf{x}}}$ are collinear, i.e., if $\det[\underline{\mathbf{x}}, \dot{\underline{\mathbf{x}}}, \ddot{\underline{\mathbf{x}}}] = 0$, then the curve has an *inflection point* at $\underline{\mathbf{x}}(t)$.[1] In projective differential geometry, one treats inflection points as special cases — then every (non-inflection) point on the curve and every higher derivative may be written as a linear combination of $\underline{\mathbf{x}}, \dot{\underline{\mathbf{x}}}, \ddot{\underline{\mathbf{x}}}$, in a fashion similar to the concept of a Frenet frame in classical differential geometry. See [26], [118].

Another interesting phenomenon arises from the observation that setting $\underline{\mathbf{y}}(t) = \rho(t)\underline{\mathbf{x}}(t)$ results in $\underline{\mathbf{y}}(t) \mathrel{\hat{=}} \underline{\mathbf{x}}(t)$ for all t. Derivatives change, however, due to application of the product rule. For the first two derivatives, we

[1] In this chapter, we shall only consider conics; these do not have inflection points.

obtain (following Geise and Nestler[2]):

$$[\underline{\mathbf{y}}, \dot{\underline{\mathbf{y}}}, \ddot{\underline{\mathbf{y}}}] = [\underline{\mathbf{x}}, \dot{\underline{\mathbf{x}}}, \ddot{\underline{\mathbf{x}}}] \begin{bmatrix} \rho & \dot{\rho} & \ddot{\rho} \\ 0 & \rho & 2\dot{\rho} \\ 0 & 0 & \rho \end{bmatrix}. \tag{4.2}$$

Similarly, let $t = \phi(s)$ represent a parameter transformation (this requires $\dot{\phi}(s) > 0$ for all s). Then we set $\underline{\mathbf{z}}(s) = \underline{\mathbf{x}}(\phi(s)) = \underline{\mathbf{x}}(t)$ and obtain, using the chain and product rules:

$$[\underline{\mathbf{z}}, \dot{\underline{\mathbf{z}}}, \ddot{\underline{\mathbf{z}}}] = [\underline{\mathbf{x}}, \dot{\underline{\mathbf{x}}}, \ddot{\underline{\mathbf{x}}}] \begin{bmatrix} 1 & 0 & 0 \\ 0 & \phi' & \phi'' \\ 0 & 0 & \phi'^2 \end{bmatrix}, \tag{4.3}$$

where the prime $'$ denotes derivatives with respect to s.

4.2 The Bernstein Form of a Conic

Let $\underline{\mathbf{b}}_0$ and $\underline{\mathbf{b}}_2$ be two distinct points on a conic Γ. Let the corresponding tangents $\underline{\mathbf{B}}_0$ and $\underline{\mathbf{B}}_2$ intersect in the point $\underline{\mathbf{b}}_1$. Let

$$\underline{\mathbf{q}}_0 = \frac{1}{2}\underline{\mathbf{b}}_0 + \frac{1}{2}\underline{\mathbf{b}}_1 \quad \text{and} \quad \underline{\mathbf{q}}_1 = \frac{1}{2}\underline{\mathbf{b}}_1 + \frac{1}{2}\underline{\mathbf{b}}_2$$

be two points on these tangents, such that $\underline{\mathbf{Q}} = \underline{\mathbf{q}}_0 \wedge \underline{\mathbf{q}}_1$ is another tangent to Γ; see Figure 4.2.[3] We call the tangent $\underline{\mathbf{Q}}$ the *shoulder tangent.*

We know that this information — two points plus three tangents — is sufficient to define the conic. Let

$$\underline{\mathbf{b}}_0^1 = \underline{\mathbf{b}}_0^1(t) = (1-t)\underline{\mathbf{b}}_0 + t\underline{\mathbf{b}}_1 \tag{4.4}$$

be a point on $\underline{\mathbf{B}}_0$. We will now locate a point $\underline{\mathbf{b}}_1^1$ on $\underline{\mathbf{B}}_2$ such that $\underline{\mathbf{T}} = \underline{\mathbf{b}}_0^1 \wedge \underline{\mathbf{b}}_1^1$ is tangent to Γ. From the four tangent theorem (3.2) we know that we must have

$$\mathrm{cr}(\underline{\mathbf{b}}_0, \underline{\mathbf{b}}_0^1, \underline{\mathbf{q}}_0, \underline{\mathbf{b}}_1) = \mathrm{cr}(\underline{\mathbf{b}}_1, \underline{\mathbf{b}}_1^1, \underline{\mathbf{q}}_1, \underline{\mathbf{b}}_2).$$

Since

$$\mathrm{cr}(\underline{\mathbf{b}}_0, \underline{\mathbf{b}}_0^1, \underline{\mathbf{q}}_0, \underline{\mathbf{b}}_1) = \frac{t}{1-t},$$

it follows that $\underline{\mathbf{b}}_1^1$ is given by

$$\underline{\mathbf{b}}_1^1 = \underline{\mathbf{b}}_1^1(t) = (1-t)\underline{\mathbf{b}}_1 + t\underline{\mathbf{b}}_2. \tag{4.5}$$

---

[2]Private communication.

[3]Keep in mind that this does not restrict the location of the $\underline{\mathbf{q}}_i$, as demonstrated in Example 1.1!

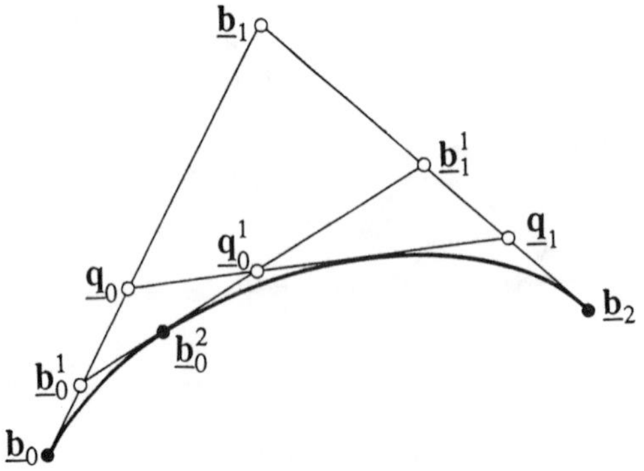

Figure 4.2. The Bézier form of conics: the conic is defined by the control polygon $\underline{\mathbf{b}}_0, \underline{\mathbf{b}}_1, \underline{\mathbf{b}}_2$ and a shoulder tangent.

If we are interested in the tangent $\underline{\mathbf{T}}$ to Γ, we are finished (for an equation of the tangent, see (4.10)). But we can go one step further and ask: what is the point of contact — we will call it $\underline{\mathbf{b}}_0^2(t)$ — between $\underline{\mathbf{T}}$ and Γ? Let us define

$$\underline{\mathbf{q}}_0^1 = \underline{\mathbf{q}}_0^1(t) = \underline{\mathbf{T}}(t) \wedge \underline{\mathbf{Q}}.$$

Since $\underline{\mathbf{q}}_0^1$ must satisfy both

$$\underline{\mathbf{q}}_0^1 = (1-\alpha)\underline{\mathbf{b}}_0^1 + \alpha\underline{\mathbf{b}}_1^1 \quad \text{and} \quad \underline{\mathbf{q}}_0^1 = (1-\beta)\underline{\mathbf{q}}_0 + \beta\underline{\mathbf{q}}_1,$$

we find that

$$\underline{\mathbf{q}}_0^1 = \frac{1}{2}\underline{\mathbf{b}}_0^1 + \frac{1}{2}\underline{\mathbf{b}}_1^1. \tag{4.6}$$

Invoking the four tangent theorem again, we see that we must have

$$\operatorname{cr}(\underline{\mathbf{b}}_0^1, \underline{\mathbf{b}}_0^2, \underline{\mathbf{q}}_0^1, \underline{\mathbf{b}}_1^1) = \frac{t}{1-t},$$

yielding

$$\underline{\mathbf{b}}_0^2(t) = (1-t)\underline{\mathbf{b}}_0^1(t) + t\underline{\mathbf{b}}_1^1(t). \tag{4.7}$$

We can now insert (4.4) and (4.5) into (4.7):

$$\underline{\mathbf{b}}_0^2(t) = (1-t)^2\underline{\mathbf{b}}_0 + 2t(1-t)\underline{\mathbf{b}}_1 + t^2\underline{\mathbf{b}}_2. \tag{4.8}$$

Defining the quadratic *Bernstein polynomials* B_i^2 as

$$B_i^2(t) = \binom{2}{i} t^i (1-t)^{2-i}; \quad i = 0, 1, 2,$$

$$\begin{array}{lll} \underline{\mathbf{b}}_0 & & \\ \underline{\mathbf{b}}_1 & \underline{\mathbf{b}}_0^1 & \\ \underline{\mathbf{b}}_2 & \underline{\mathbf{b}}_1^1 & \underline{\mathbf{b}}_0^2 \end{array} \qquad \begin{array}{cccccc} (1-t) & \bullet & & & & \\ & & \searrow & & & \\ t & \bullet & \rightarrow & + & \rightarrow & \bullet \end{array}$$

$$\begin{array}{ccc} \begin{bmatrix} 0 \\ 0 \\ 1 \end{bmatrix} & & \\ \begin{bmatrix} 0 \\ 2 \\ 1 \end{bmatrix} & \begin{bmatrix} 0 \\ 1 \\ 1 \end{bmatrix} & \\ \begin{bmatrix} 8 \\ 2 \\ 5 \end{bmatrix} & \begin{bmatrix} 4 \\ 2 \\ 3 \end{bmatrix} & \begin{bmatrix} 2 \\ \frac{3}{2} \\ 2 \end{bmatrix} \end{array}$$

Example 4.1. The triangular scheme for the conic de Casteljau algorithm. Top left: the overall scheme; top right: the rule to generate it. Bottom: a numerical example for $t = \frac{1}{2}$.

we can write our conic as a *projective quadratic Bézier curve:*

$$\underline{\mathbf{b}}_0^2(t) = \sum_{i=0}^{2} \underline{\mathbf{b}}_i B_i^2(t). \tag{4.9}$$

This form of the conic is a *parametric quadratic curve.* As the parameter t traces out all values between $+\infty$ and $-\infty$, the point $\underline{\mathbf{b}}_0^2(t)$ traces out all of the conic. We also use the term "quadratic Bézier curve." This terminology will be explained in Chapter 7.

Our recursive algorithm, called the *de Casteljau algorithm,* is conveniently written in a triangular arrangement as shown in Example 4.1.

We should note that, although we worked in the context of line conics — using the four tangent theorem — we finally arrived at the equation of a *point* on the conic. For the corresponding tangent, $\underline{\mathbf{T}}$, we write

$$\begin{aligned} \underline{\mathbf{T}}(t) &= \underline{\mathbf{b}}_0^1 \wedge \underline{\mathbf{b}}_1^1 \\ &= [(1-t)\underline{\mathbf{b}}_0 + t\underline{\mathbf{b}}_1] \wedge [(1-t)\underline{\mathbf{b}}_1 + t\underline{\mathbf{b}}_2] \\ &= (1-t)^2[\underline{\mathbf{b}}_0 \wedge \underline{\mathbf{b}}_1] + t(1-t)[\underline{\mathbf{b}}_0 \wedge \underline{\mathbf{b}}_2] + t(1-t)[\underline{\mathbf{b}}_1 \wedge \underline{\mathbf{b}}_1] + t^2[\underline{\mathbf{b}}_1 \wedge \underline{\mathbf{b}}_2]. \end{aligned}$$

We observe that $\underline{\mathbf{b}}_1 \wedge \underline{\mathbf{b}}_1 = \mathbf{0}$ and arrive at

$$\underline{\mathbf{T}}(t) = (1-t)^2\underline{\mathbf{B}}_0 + t(1-t)\underline{\mathbf{B}}_1 + t^2\underline{\mathbf{B}}_2 \tag{4.10}$$

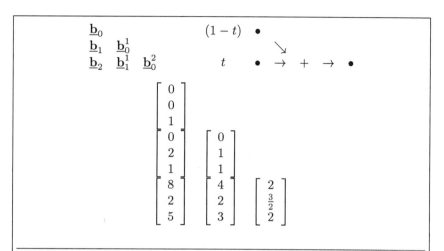

for the equation of the tangent, expressed in terms of the control polygon legs

$$\underline{\mathbf{B}}_0 = \underline{\mathbf{b}}_0 \wedge \underline{\mathbf{b}}_1, \quad \underline{\mathbf{B}}_1 = \underline{\mathbf{b}}_0 \wedge \underline{\mathbf{b}}_2, \quad \underline{\mathbf{B}}_2 = \underline{\mathbf{b}}_1 \wedge \underline{\mathbf{b}}_2.$$

For the special case $t = \frac{1}{2}$, we obtain

$$\underline{\mathbf{q}}_0 \wedge \underline{\mathbf{q}}_1 = \underline{\mathbf{B}}_0 + \underline{\mathbf{B}}_1 + \underline{\mathbf{B}}_2.$$

4.3 Parametric Point conics

Our above development was in the context of line conics. Clearly, we should also be able to derive parametric point conics. We would then be given two points $\underline{\mathbf{b}}_0$ and $\underline{\mathbf{b}}_2$ and the respective tangents $\underline{\mathbf{B}}_0$ and $\underline{\mathbf{B}}_2$; see Figure 4.3. In addition, another point $\underline{\mathbf{q}}$ on the conic would be given, defining two lines $\underline{\mathbf{Q}}_0 = \underline{\mathbf{q}} \wedge \underline{\mathbf{b}}_0$ and $\underline{\mathbf{Q}}_1 = \underline{\mathbf{q}} \wedge \underline{\mathbf{b}}_2$.

We need three pairs of lines from the pencils through $\underline{\mathbf{b}}_0$ and $\underline{\mathbf{b}}_2$, thus defining a projectivity Φ between the two pencils. We select, with the notation from Figure 4.3,

$$\underline{\mathbf{B}}_2 = \Phi\underline{\mathbf{B}}_1, \quad \underline{\mathbf{Q}}_1 = \Phi\underline{\mathbf{Q}}_0, \quad \underline{\mathbf{B}}_1 = \Phi\underline{\mathbf{B}}_0.$$

Again, we set $\underline{\mathbf{Q}}_0 = \frac{1}{2}\underline{\mathbf{B}}_0 + \frac{1}{2}\underline{\mathbf{B}}_1$ and $\underline{\mathbf{Q}}_1 = \frac{1}{2}\underline{\mathbf{B}}_1 + \frac{1}{2}\underline{\mathbf{B}}_2$. Selecting an arbitrary line

$$\underline{\mathbf{B}}_0^1(t) = (1-t)\underline{\mathbf{B}}_0 + t\underline{\mathbf{B}}_1$$

in the pencil through $\underline{\mathbf{b}}_0$, and utilizing the four point theorem (3.3), we find its image under Φ to be $\underline{\mathbf{B}}_1^1(t) = (1-t)\underline{\mathbf{B}}_1 + t\underline{\mathbf{B}}_2$.

We may set $\underline{\mathbf{Q}}_0^1 = \frac{1}{2}\underline{\mathbf{B}}_0^1 + \frac{1}{2}\underline{\mathbf{B}}_1^1$ in analogy to (4.6). We can now apply the four point theorem to the pencil of lines through $\underline{\mathbf{B}}_0^1 \wedge \underline{\mathbf{B}}_1^1$, and find that

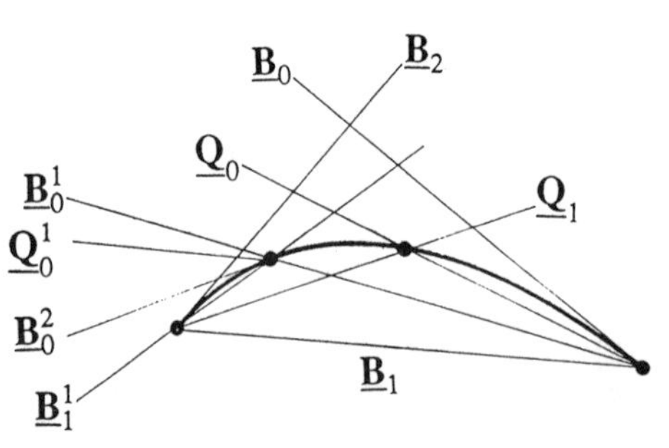

Figure 4.3. Parametric point conics: two pencils and an additional point are given.

the tangent $\underline{\mathbf{B}}_0^2(t)$ to the conic must satisfy $\underline{\mathbf{B}}_0^2 = (1-t)\underline{\mathbf{B}}_0^1 + t\underline{\mathbf{B}}_1^1$, resulting in

$$\underline{\mathbf{B}}_0^2(t) = \sum_{i=0}^{2} \underline{\mathbf{B}}_i B_i^2(t), \tag{4.11}$$

which is of course the dual of (4.9). We will encounter this type of equation again in Section 7.11.

This is the expression for the tangent to a point conic; the corresponding point $\underline{\mathbf{p}}(t)$ is found, by complete analogy to (4.10):

$$\underline{\mathbf{p}}(t) = (1-t)^2\underline{\mathbf{b}}_0 + t(1-t)\underline{\mathbf{b}}_1 + t^2\underline{\mathbf{b}}_2. \tag{4.12}$$

4.4 Derivatives

Now for the derivative of a projective conic in Bézier form: it is given by

$$\dot{\mathbf{b}}(t) = 2(1-t)\Delta\underline{\mathbf{b}}_0 + 2t\Delta\underline{\mathbf{b}}_1,$$

or, equivalently,

$$\dot{\mathbf{b}}(t) \hat{=} (1-t)\Delta\underline{\mathbf{b}}_0 + t\Delta\underline{\mathbf{b}}_1, \tag{4.13}$$

where Δ is the *forward difference operator:*

$$\Delta\underline{\mathbf{b}}_i = \underline{\mathbf{b}}_{i+1} - \underline{\mathbf{b}}_i.$$

Since both $\Delta\underline{\mathbf{b}}_0$ and $\Delta\underline{\mathbf{b}}_1$ are points, it follows that (4.13) describes a line, and we have that *all derivatives of a conic are collinear.* This is to be expected, of course, since the derivative of a quadratic is linear.

Where is this line on which all conic derivatives exist? Let us rewrite (4.9) as

$$\underline{\mathbf{b}}(t) = \underline{\mathbf{b}}_0 + t\Delta\underline{\mathbf{b}}_0 + \frac{1}{2}t^2\Delta^2\underline{\mathbf{b}}_0, \tag{4.14}$$

with

$$\Delta^2\underline{\mathbf{b}}_0 = \Delta\underline{\mathbf{b}}_1 - \Delta\underline{\mathbf{b}}_0. \tag{4.15}$$

We now see that

$$\underline{\mathbf{b}}(\infty) := \lim_{t\to\infty} \underline{\mathbf{b}}(t) \hat{=} \Delta^2\underline{\mathbf{b}}_0. \tag{4.16}$$

Using the same limit argument again, we obtain

$$\underline{\mathbf{b}}(\infty) \hat{=} \dot{\underline{\mathbf{b}}}(\infty) \hat{=} \ddot{\underline{\mathbf{b}}}(\infty), \tag{4.17}$$

where $\ddot{\underline{\mathbf{b}}}$ stands for the second derivative. Note that the second derivative is a point on the conic!

Since $\Delta\underline{\mathbf{b}}_0, \Delta^2\underline{\mathbf{b}}_0$, and $\Delta\underline{\mathbf{b}}_1$ are collinear points by virtue of (4.15), it follows that all first derivatives to $\underline{\mathbf{b}}$ lie on the tangent corresponding to $t = \infty$:

$$\dot{\underline{\mathbf{b}}}(t) = \underline{\mathbf{T}}(\infty) \wedge \underline{\mathbf{T}}(t). \tag{4.18}$$

For an illustration, see Figure 4.4.

That figure suggests another identity:

$$\det[\underline{\mathbf{b}}_1, \underline{\mathbf{b}}(\frac{1}{2}), \underline{\mathbf{b}}(\infty)] = 0, \tag{4.19}$$

i.e., the three points $\underline{\mathbf{b}}_1, \underline{\mathbf{b}}(\frac{1}{2}), \underline{\mathbf{b}}(\infty)$ are collinear. To prove this fact, we verify the simple identity

$$\underline{\mathbf{b}}(\infty) - \underline{\mathbf{b}}(\frac{1}{2}) \hat{=} 4\underline{\mathbf{b}}_1.$$

The endpoints of a conic segment deserve special attention: their derivatives are given by

$$\dot{\underline{\mathbf{b}}}(0) = 2\Delta\underline{\mathbf{b}}_0 \quad \text{and} \quad \dot{\underline{\mathbf{b}}}(1) = 2\Delta\underline{\mathbf{b}}_1. \tag{4.20}$$

Note that the four points $\underline{\mathbf{b}}_0, \underline{\mathbf{q}}_0, \underline{\mathbf{b}}_1, \dot{\underline{\mathbf{b}}}(0)$ are harmonic, as are $\underline{\mathbf{b}}_1, \underline{\mathbf{q}}_1, \underline{\mathbf{b}}_2, \dot{\underline{\mathbf{b}}}(1)$.

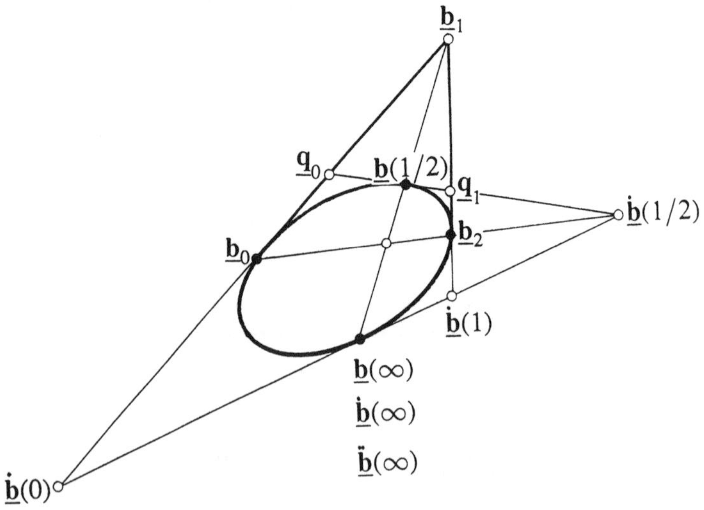

Figure 4.4. Tangents and derivatives: several interesting cases.

4.5 The Implicit Form

The parametric form of a conic offers a means to compute points on it: given a parameter value t, we can find the corresponding point on the conic. On the other hand, it offers no solution to the following problem: given a point $\underline{\mathbf{x}}$, is it on the conic or not? The implicit form will help answer this question. We shall first derive the implicit form for a special case of parametric conics, then we will take it to the general case.

Let us assign the following coordinates to the defining points of the conic Γ:

$$\underline{\mathbf{b}}_0 = \begin{bmatrix} 0 \\ 0 \\ 1 \end{bmatrix}, \; \underline{\mathbf{b}}_1 = \begin{bmatrix} 0 \\ 1 \\ 0 \end{bmatrix}, \; \underline{\mathbf{b}}_2 = \begin{bmatrix} 1 \\ 0 \\ 0 \end{bmatrix}, \; \underline{\mathbf{q}}_0 = \frac{1}{2}\begin{bmatrix} 0 \\ 1 \\ 1 \end{bmatrix}, \; \underline{\mathbf{q}}_1 = \frac{1}{2}\begin{bmatrix} 1 \\ 1 \\ 0 \end{bmatrix}.$$

Then Γ takes on the simple form

$$\underline{\mathbf{x}} = \underline{\mathbf{x}}(t) = \begin{bmatrix} t^2 \\ 2t(1-t) \\ (1-t)^2 \end{bmatrix}, \tag{4.21}$$

from which we deduce the identity

$$y^2 = 4xz. \tag{4.22}$$

This expression — the implicit form of a conic — does not involve the parameter value t anymore! It holds for any point on the conic, regardless of its parameter value.

We can rewrite (4.22) in matrix form:

$$\underline{\mathbf{x}}^{\mathrm{T}} A \underline{\mathbf{x}} = 0, \tag{4.23}$$

with the symmetric matrix

$$A = \begin{bmatrix} 0 & 0 & -2 \\ 0 & 1 & 0 \\ -2 & 0 & 0 \end{bmatrix}.$$

Now we can decide if a given point is on Γ just by checking if its coordinates satisfy (4.23) or not.

Let us define the line

$$\underline{\mathbf{T}} = \underline{\mathbf{x}}^{\mathrm{T}} A. \tag{4.24}$$

Then $\underline{\mathbf{T}}\underline{\mathbf{x}} = 0$ by (4.23), and so $\underline{\mathbf{x}}$ lies on $\underline{\mathbf{T}}$. In terms of the parameter t, we can express $\underline{\mathbf{T}}$ as

$$\underline{\mathbf{T}} = 2[-(1-t)^2, t(1-t), -t^2].$$

In our special coordinate system, we have

$$\underline{\mathbf{b}}_0^1 = \begin{bmatrix} 0 \\ t \\ 1-t \end{bmatrix} \quad \text{and} \quad \underline{\mathbf{b}}_1^1 = \begin{bmatrix} t \\ 1-t \\ 0 \end{bmatrix}.$$

It is now easy to check that $\underline{\mathbf{Tb}}_0^1 = \underline{\mathbf{Tb}}_1^1 = 0$, i.e., the three collinear points $\underline{\mathbf{b}}_0^1, \underline{\mathbf{b}}_0^2, \underline{\mathbf{b}}_1^1$ all are on $\underline{\mathbf{T}}$. Thus $\underline{\mathbf{T}}$ is the tangent to Γ at $\underline{\mathbf{x}}$!

So far, our implicit approach lacks generality: it seems to rely heavily on the special coordinate system that we have used. We will now show that this is not so. Let $\underline{\mathbf{a}}, \underline{\mathbf{b}}, \underline{\mathbf{c}}, \underline{\mathbf{d}}$ form a projective reference frame. With respect to it, the point $\underline{\mathbf{x}}$ has coordinates $\hat{\underline{\mathbf{x}}}$, given by a coordinate transformation $\hat{\underline{\mathbf{x}}} = M\underline{\mathbf{x}}$, where M is a nonsingular 3×3 matrix. Then the implicit form of our conic Γ becomes

$$\underline{\mathbf{x}}^{\mathrm{T}} M^{\mathrm{T}} A M \underline{\mathbf{x}} = 0.$$

Since $M^{\mathrm{T}}AM$ is again a nonsingular symmetric 3×3 matrix, we see that no matter what coordinate system we use, a conic can always be written as $\underline{\mathbf{x}}^{\mathrm{T}}B\underline{\mathbf{x}} = 0$, with B a nonsingular symmetric 3×3 matrix. If $\underline{\mathbf{x}}$ is a point on this conic, then $\underline{\mathbf{x}}^{\mathrm{T}}A$ is the corresponding tangent.

We may use the implicit form to talk about points "inside" and "outside" the conic: define all points that satisfy $\underline{\mathbf{x}}^{\mathrm{T}}A\underline{\mathbf{x}} < 0$ to be inside the conic, and all those with $\underline{\mathbf{x}}^{\mathrm{T}}A\underline{\mathbf{x}} > 0$ to be outside.[4] This definition is invariant under projective maps and allows us to split the projective plane into two separate parts.

The implicit form of an affine conic is given by (4.12) in the Problems section.

We have shown in this section how to obtain the implicit form of a conic once we have a parametric representation of it. How about the other way around: can we go from the implicit form to the parametric form? We can, but not in a unique way. The implicit form does not single out any points on the conic, whereas the parametric one does: we have to specify two points, $\underline{\mathbf{b}}_0$ and $\underline{\mathbf{b}}_2$ on the conic, together with their tangents, plus a third tangent. So if we are given the implicit form, we first have to determine two points on it, and then find their tangents. This can be done using (4.24). The interplay between parametric and implicit forms, also for higher degree curves, has recently attracted some attention. We list: [1], [2], [3], [62], [63], [103], [106].

[4] Of course, we could define this the other way around as well!

4.6 Implicit Interpolating Conics

We know that five points determine a conic. What is its implicit form? To answer that question, assume that the points are $\underline{\mathbf{x}}_1, \ldots, \underline{\mathbf{x}}_5$. Any point $\underline{\mathbf{x}}$ on the desired conic must satisfy

$$\begin{vmatrix} x^2 & y^2 & z^2 & xy & xz & yz \\ x_1^2 & y_1^2 & z_1^2 & x_1y_1 & x_1z_1 & y_1z_1 \\ x_2^2 & y_2^2 & z_2^2 & x_2y_2 & x_2z_2 & y_2z_2 \\ x_3^2 & y_3^2 & z_3^3 & x_3y_3 & x_3z_3 & y_3z_3 \\ x_4^2 & y_4^2 & z_4^2 & x_4y_4 & x_4z_4 & y_4z_4 \\ x_5^2 & y_5^2 & z_5^2 & x_5y_5 & x_5z_5 & y_5z_5 \end{vmatrix} = 0. \tag{4.25}$$

The above expression is quadratic in x, y, z; thus we can bring it into the form (4.5). Since it vanishes for $\underline{\mathbf{x}} = \underline{\mathbf{x}}_i; i = 1, \ldots, 5$, it interpolates to the given points.

Five *affine* points $\mathbf{x}_1, \ldots, \mathbf{x}_5$ also determine a conic; its implicit form is given by

$$\begin{vmatrix} x^2 & xy & y^2 & x & y & 1 \\ x_1^2 & x_1y_1 & y_1^2 & x_1 & y_1 & 1 \\ x_2^2 & x_2y_2 & y_2^2 & x_2 & y_2 & 1 \\ x_3^2 & x_3y_3 & y_3^2 & x_3 & y_3 & 1 \\ x_4^2 & x_4y_4 & y_4^2 & x_4 & y_4 & 1 \\ x_5^2 & x_5y_5 & y_5^2 & x_5 & y_5 & 1 \end{vmatrix} = 0, \tag{4.26}$$

which is obtained from (4.25) by setting

$$\mathbf{x} = \begin{bmatrix} x \\ y \end{bmatrix} \leftarrow \begin{bmatrix} x/z \\ y/z \\ 1 \end{bmatrix}.$$

Left of the arrow, we have affine coordinates, while there are projective ones right of it!

It should be noted that process of finding an interpolating conic may be numerically unstable: if the five points are "close," they do not convey much information about the conic in an intuitive sense. If we use (4.25) or (4.26) to check if a point is on a conic that is determined by five points, these expressions will in general not evaluate to zero, but to some small number. Thus we cannot check our determinants for being zero, but rather against a small tolerance ϵ. If the given points were too close to each other, many points not on the intended conic will satisfy the specified tolerance, and thus be labeled "on the conic."

4.7 Parametric Interpolating Conics

Parametric conics (4.9) may be used to solve the following interpolation problem: given four points $\underline{\mathbf{x}}_0, \ldots, \underline{\mathbf{x}}_3$ and corresponding parameter values $t_0, \ldots, t_3$, find a conic $\underline{\mathbf{c}}(t)$ such that $\underline{\mathbf{c}}(t_i) \hat{=} \underline{\mathbf{x}}_i$. Figure 4.5 illustrates.

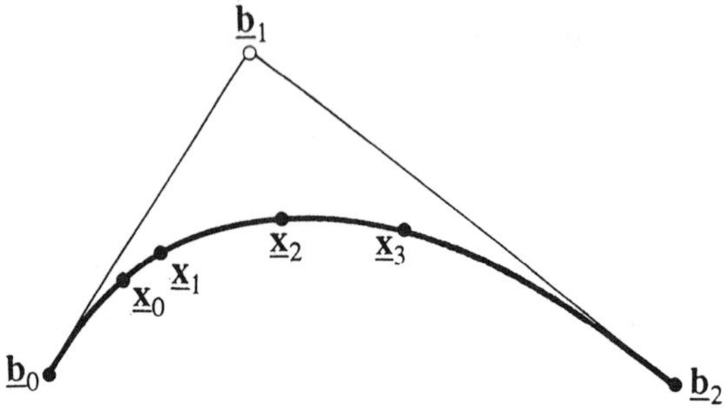

Figure 4.5. Parametric conic interpolation: the given data together with the unknown control polygon.

Since we will be using linear systems, we first rewrite (4.9) as

$$\underline{\mathbf{c}}(t) = [B_0^2(t), B_1^2(t), B_2^2(t)] \begin{bmatrix} \underline{\mathbf{b}}_0 \\ \underline{\mathbf{b}}_1 \\ \underline{\mathbf{b}}_2 \end{bmatrix}. \tag{4.27}$$

Note that the column on the right has three points as its elements. We now write down the interpolation conditions for the first three data points:

$$\begin{bmatrix} z_0\underline{\mathbf{x}}_0 \\ z_1\underline{\mathbf{x}}_1 \\ z_2\underline{\mathbf{x}}_2 \end{bmatrix} = \begin{bmatrix} B_0^2(t_0) & B_1^2(t_0) & B_2^2(t_0) \\ B_0^2(t_1) & B_1^2(t_1) & B_2^2(t_1) \\ B_0^2(t_2) & B_1^2(t_2) & B_2^2(t_2) \end{bmatrix} \begin{bmatrix} \underline{\mathbf{b}}_0 \\ \underline{\mathbf{b}}_1 \\ \underline{\mathbf{b}}_2 \end{bmatrix}. \tag{4.28}$$

The factors z_i are — unknown — scaling factors. We can rewrite (4.28) by naming the involved vectors[5] and matrix:

$$\boxed{\mathbf{x}} = B\boxed{\mathbf{b}}.$$

This has the solution $\boxed{\mathbf{b}} = B^{-1}\boxed{\mathbf{x}}$. The problem is, $\boxed{\mathbf{x}}$ involves the unknown z_i, and they still have to be found. Now we make use of the

[5] Column vectors that have points as their elements are marked by boxes. Keep in mind that this notation is shorthand for *three* linear systems, one for each of the $x-, y-, z-$ components!

fourth point, $\underline{\mathbf{x}}_3$. We may set $z = 1$ for one point without loss of generality, and we do this for $\underline{\mathbf{x}}_3$. Then (4.27) becomes

$$\underline{\mathbf{x}}_3 = \underline{\mathbf{c}}(t_3) = [B_0^2(t_3), B_1^2(t_3), B_2^2(t_3)] \begin{bmatrix} \underline{\mathbf{b}}_0 \\ \underline{\mathbf{b}}_1 \\ \underline{\mathbf{b}}_2 \end{bmatrix}. \tag{4.29}$$

Abbreviating this as $\underline{\mathbf{x}}_3 = B_3\boxed{\mathbf{b}}$, we have

$$\underline{\mathbf{x}}_3 = B_3 B^{-1}\boxed{\mathbf{x}}.$$

This constitutes a 3×3 linear system for the z_i, which are "hidden" in $\boxed{\mathbf{x}}$. There is no guarantee, of course, that the z_i will turn out to be positve — thus if we project into affine space (see Chapter 5), we may encounter asymptotes.

Normally, we say that a conic is determined by five points on it. In that case, no parameter values are used. If we introduce parameters that are attached to the data points, then four points are sufficient!

4.8 Blossoms and Polars

We now introduce a generalization of the de Casteljau algorithm. It will unify several of the previously developed concepts.

The de Casteljau algorithm for conics is a two-level procedure: at the first level, we interpolate (twice) with respect to t to obtain the two $\underline{\mathbf{b}}_i^1(t)$. At the second level, we interpolate with respect to t again, obtaining $\underline{\mathbf{b}}_0^2(t)$. Our generalization is to allow two *different* parameters at the two levels. While the "standard" de Casteljau algorithm yields a point that depends on one variable: $\underline{\mathbf{b}}_0^2(t)$, we now obtain a point that depends on two variables. We denote it by $\underline{\mathbf{b}}[s,t]$, the two variables being s and t. Of course, if $s = t$, we have $\underline{\mathbf{b}}[t,t] = \underline{\mathbf{b}}_0^2(t)$. We call $\underline{\mathbf{b}}[s,t]$ the *blossom* of the conic; this terminology was introduced by L. Ramshaw in 1987; see [96] or [97].

The blossom $\underline{\mathbf{b}}[s,t]$ is a *symmetric function*: this means that $\underline{\mathbf{b}}[s,t] = \underline{\mathbf{b}}[t,s]$. This is verified by a straightforward exercise in algebra.

Just to get used to the idea of blossoms, let us compute a few special values. We already know $\underline{\mathbf{b}}[0,0] = \underline{\mathbf{b}}_0$ and $\underline{\mathbf{b}}[1,1] = \underline{\mathbf{b}}_2$. Next, let us try $\underline{\mathbf{b}}[0,1]$. We get the following scheme:

$$\begin{array}{lll} \underline{\mathbf{b}}_0 & & \\ \underline{\mathbf{b}}_1 & \underline{\mathbf{b}}_0 & \\ \underline{\mathbf{b}}_2 & \underline{\mathbf{b}}_1 & \underline{\mathbf{b}}_1 = \underline{\mathbf{b}}[0,1], \end{array}$$

where we have used $s = 0$ first and $t = 1$ second – the reader is invited to verify that the same result is obtained by using $t = 1$ first and then $s = 0$. Thus all conic control points can be written as blossom values!

Similarly, we can show that $\underline{\mathbf{b}}[0,t] = \underline{\mathbf{b}}_0^1(t)$ and $\underline{\mathbf{b}}_1^1(t) = \underline{\mathbf{b}}[1,t]$. Thus we can reformulate the de Casteljau scheme in blossom form:

$$\begin{array}{lll} \underline{\mathbf{b}}_0 & & \\ \underline{\mathbf{b}}_1 & \underline{\mathbf{b}}_0^1(t) & \\ \underline{\mathbf{b}}_2 & \underline{\mathbf{b}}_1^1(t) & \underline{\mathbf{b}}_0^2(t) \end{array} \quad = \quad \begin{array}{lll} \underline{\mathbf{b}}[0,0] & & \\ \underline{\mathbf{b}}[0,1] & \underline{\mathbf{b}}[0,t] & \\ \underline{\mathbf{b}}[1,1] & \underline{\mathbf{b}}[1,t] & \underline{\mathbf{b}}[t,t]. \end{array}$$

Figure 4.6 shows a conic with several blossom values; we will discuss those now. First, we observe that if r is a fixed parameter, then $\underline{\mathbf{b}}[r,t]$ traces out all points on the tangent $\underline{\mathbf{T}}(r)$ at $\underline{\mathbf{b}}[r,r]$. In particular, $\underline{\mathbf{b}}[r,t]$ is the intersection of the tangents at $\underline{\mathbf{b}}[r,r]$ and $\underline{\mathbf{b}}[t,t]$:

$$\underline{\mathbf{b}}[r,t] = \underline{\mathbf{T}}(r) \wedge \underline{\mathbf{T}}(t), \tag{4.30}$$

which gives a geometric argument for the symmetry of the blossom function.

In classical geometry, one also sees the definition that the point $\underline{\mathbf{b}}[r,t]$ is the *pole* to the *polar*

$$\underline{\mathbf{B}}[r,t] = \underline{\mathbf{b}}[r,r] \wedge \underline{\mathbf{b}}[t,t],$$

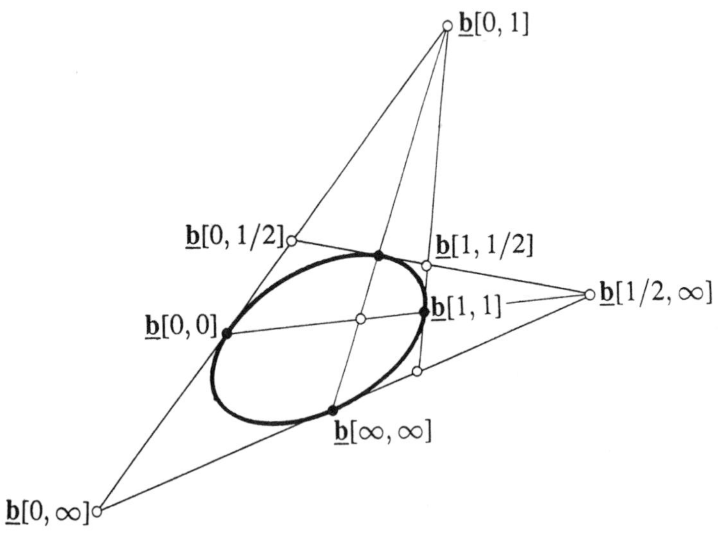

Figure 4.6. Conic blossoms: several blossom values.

as illustrated in Figure 4.7. Note that not every point has a corresponding polar — we call those points interior to the conic. If we wanted to be able to define polars for *every* point, we would have to work in a projective plane that is defined over complex numbers.

For the case $r = t$, this becomes

$$\underline{\mathbf{B}}[r,r] = \underline{\mathbf{T}}(r),$$

i.e., the polar of a point on a conic is its tangent.

We can also write derivatives of a conic in blossom form:

$$\dot{\underline{\mathbf{b}}}(t) \mathrel{\hat{=}} \underline{\mathbf{b}}[\infty, t], \ddot{\underline{\mathbf{b}}}(t) \mathrel{\hat{=}} \underline{\mathbf{b}}[\infty, \infty]. \tag{4.31}$$

Thus taking derivatives amounts to evaluating at $t = \infty$, and all derivatives to a conic are on the line $\underline{\mathbf{B}}[\infty, t]$.

Next, we show the following identity:

$$\mathrm{cr}(\underline{\mathbf{b}}[0,t], \underline{\mathbf{b}}[t,t], \underline{\mathbf{b}}[1,t], \underline{\mathbf{b}}[\infty,t]) = \frac{1}{2}, \tag{4.32}$$

i.e., the four points with parameter values $0, t, 1, \infty$ on a tangent $\underline{\mathbf{T}}(t)$ are harmonic. Referring to Figure 4.8, we see that the four points $\underline{\mathbf{b}}[0,t]$, $\underline{\mathbf{b}}[0,1]$, $\underline{\mathbf{b}}[1,t]$, $\underline{\mathbf{x}}$ form a quadrilateral such that $\underline{\mathbf{b}}[0,0]$ and $\underline{\mathbf{b}}[1,1]$ are the intersections of opposite edges. Note that $\underline{\mathbf{b}}[0,1]$, $\underline{\mathbf{b}}[t,t]$, $\underline{\mathbf{x}}$ are collinear as a consequence of Pascal's theorem; see Section 3.4.

A more general definition of poles and polars also exists (see, e.g., [111]): let $\underline{\mathbf{p}}$ be an arbitrary point. Draw any line through $\underline{\mathbf{p}}$ and record its two

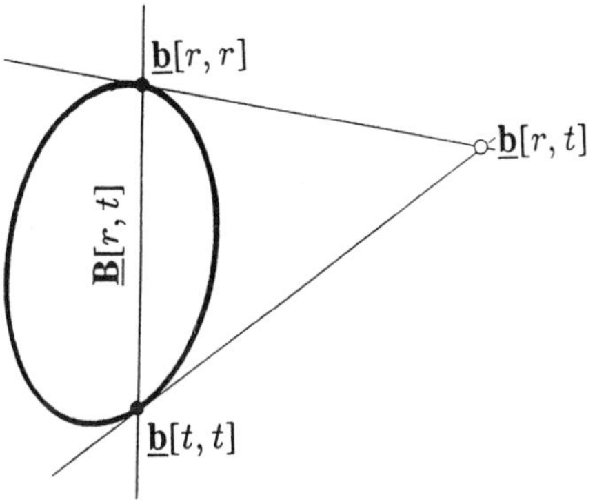

Figure 4.7. Pole and polar: a point outside the conic is related to a secant of the conic, namely its polar.

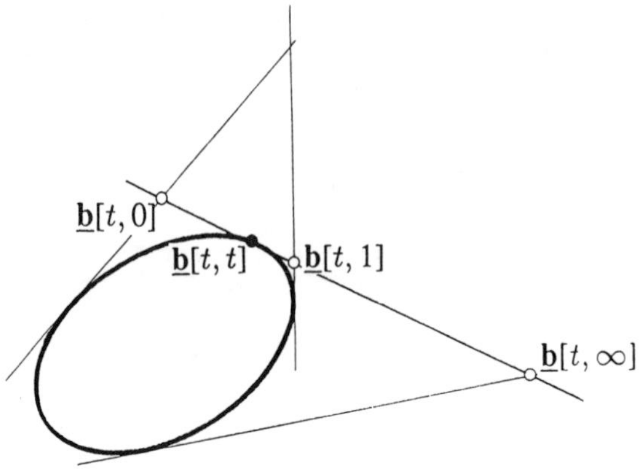

Figure 4.8. Blossoms: the generation of harmonic points.

intersections $\underline{\mathbf{p}}_1, \underline{\mathbf{p}}_2$ with a given conic.[6] The locus of all points $\underline{\mathbf{p}}^*$ such that $\underline{\mathbf{p}}_1, \underline{\mathbf{p}}^*, \underline{\mathbf{p}}_2, \underline{\mathbf{p}}$ are harmonic is a line, namely the polar with respect to $\underline{\mathbf{p}}$. Our definition of poles and polars also generates harmonic points, as was outlined in Section 4.4.

4.9 Reparametrization

In our definition of parametric conics, we had picked a tangent through two points $\underline{\mathbf{q}}_0$ and $\underline{\mathbf{q}}_1$. We then assumed that all point coordinates were rescaled such that

$$\underline{\mathbf{q}}_0 = \frac{1}{2}\underline{\mathbf{b}}_0 + \frac{1}{2}\underline{\mathbf{b}}_1 \quad \text{and} \quad \underline{\mathbf{q}}_1 = \frac{1}{2}\underline{\mathbf{b}}_1 + \frac{1}{2}\underline{\mathbf{b}}_2. \tag{4.33}$$

Let $\hat{\underline{\mathbf{q}}}_0 \wedge \hat{\underline{\mathbf{q}}}_1$ define yet another tangent to our conic, as shown in Figure 4.9. Suppose that, with (4.33) still holding, we have $\mathrm{cr}(\underline{\mathbf{b}}_0, \hat{\underline{\mathbf{q}}}_0, \underline{\mathbf{q}}_0, \underline{\mathbf{b}}_1) = c$. Because of the four tangent theorem, the four points $\underline{\mathbf{b}}_1, \hat{\underline{\mathbf{q}}}_1, \underline{\mathbf{q}}_1, \underline{\mathbf{b}}_2$ must also be in cross ratio c.

Following a similar development to that in Section 4.2, we arrive at

$$\underline{\mathbf{b}}(s) = (1-s)^2\underline{\mathbf{b}}_0 + 2cs(1-s)\underline{\mathbf{b}}_1 + c^2s^2\underline{\mathbf{b}}_2 \tag{4.34}$$

as the equation of the conic with $\hat{\underline{\mathbf{q}}}_0 \wedge \hat{\underline{\mathbf{q}}}_1$ as the shoulder tangent corresponding to $s = \frac{1}{2}$. The choice of a new shoulder tangent can be viewed as

[6] In order to always have two intersections, one might have to resort to working with the complex projective plane.

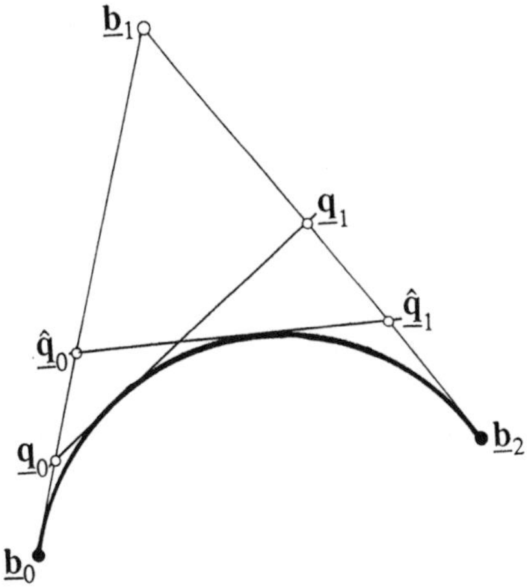

Figure 4.9. Conic reparametrization: a new tangent can be given the role of shoulder tangent.

a *reparametrization* of the conic. Since

$$\hat{\underline{\mathbf{q}}}_0 \,\hat{=}\, \underline{\mathbf{b}}_0 + c\underline{\mathbf{b}}_1,$$

the new parameter s is related to the old parameter t by

$$t = 0 \leftrightarrow s = 0, \quad t = \frac{1}{2} \leftrightarrow s = r, \quad t = 1 \leftrightarrow s = 1, \tag{4.35}$$

with $r = c/(1+c)$ since then $\operatorname{cr}(0, r, \frac{1}{2}, 1) = c$. Thus there exists a Moebius transformation between the parameters s and t. It is characterized by

$$\operatorname{cr}(0, t, \frac{1}{2}, 1) = \operatorname{cr}(0, s, r, 1)$$

and thus

$$s = \frac{ct}{1 + (c-1)t}. \tag{4.36}$$

It is important to note that, although the shape of the conic is unchanged by such a reparametrization, its derivatives change: we now have

$$\dot{\underline{\mathbf{b}}}(0) = 2c\underline{\mathbf{b}}_1 - 2\underline{\mathbf{b}}_0 \quad \text{and} \quad \dot{\underline{\mathbf{b}}}(1) = 2c^2\underline{\mathbf{b}}_2 - 2c\underline{\mathbf{b}}_1. \tag{4.37}$$

Using the concept of reparametrization, we may bring conics into *standard form*: assuming $z_0 = 1$ (thus excluding the case $z_0 = 0$!), we can set

$$c = \sqrt{\frac{1}{z_2}} \tag{4.38}$$

and can thus achieve $z_2 = 1$ as well. Of course, this only works if $z_2 > 0$ – thus not all quadratics can be brought into standard form. In Chapter 5, we will be concerned mostly with conics for which all three z_i are positive; then the normal form is always possible.

4.10 The Complementary Segment

Let us consider the particular reparametrization arising from selecting $c = -1$ (or, equivalently, $r = \infty$) in (4.36). Then the three points $\underline{\mathbf{b}}(t), \underline{\mathbf{b}}(s), \underline{\mathbf{b}}_1$ are collinear, as follows from

$$\begin{aligned}
\det[\underline{\mathbf{b}}(t),\ \underline{\mathbf{b}}(s),\ \underline{\mathbf{b}}_1] & \\
&= \det[\underline{\mathbf{b}}_0 B_0^2(t) + \underline{\mathbf{b}}_1 B_1^2(t) + \underline{\mathbf{b}}_2 B_2^2(t),\ \underline{\mathbf{b}}_0 B_0^2(t) \\
&\quad -\underline{\mathbf{b}}_1 B_1^2(t) + \underline{\mathbf{b}}_2 B_2^2(t),\ \underline{\mathbf{b}}_1] \\
&= \det[\underline{\mathbf{b}}_1 (B_1^2(t) - B_1^2(s)),\ \underline{\mathbf{b}}(s),\ \underline{\mathbf{b}}_1] \\
&= 0,
\end{aligned}$$

assuming standard form for both conics. Referring to Figure 4.10, we see that $\underline{\mathbf{b}}(s), 0 \leq s \leq 1$, traces out the part of the conic that is "missed" by $\underline{\mathbf{b}}(t), 0 \leq t \leq 1$.

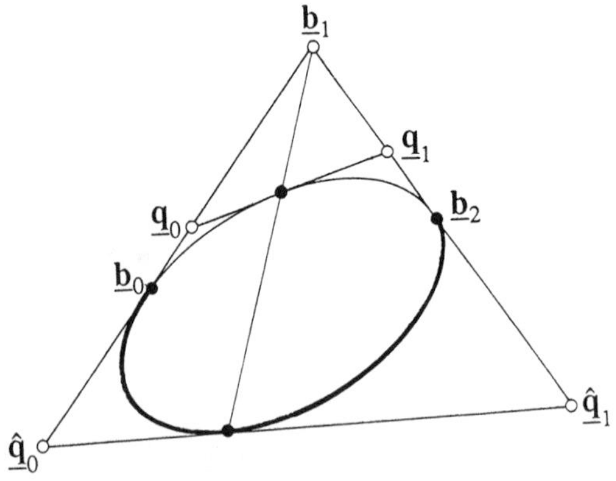

Figure 4.10. The complementary segment: reversing the weight of the middle control point generates the "other" part of the conic.

A different way of seeing this is provided by the following argument: instead of the tangent $\underline{\mathbf{T}}(\frac{1}{2}) = \underline{\mathbf{q}}_0 \wedge \underline{\mathbf{q}}_1$, we might have prescribed $\underline{\mathbf{T}}(\infty)$ as the shoulder tangent. The roles of $\underline{\mathbf{T}}(\frac{1}{2})$ and $\underline{\mathbf{T}}(\infty)$ are then reversed.

4.11 Parametric Affine Conics

If we project from $\mathbb{P}^2$ into the extended affine plane, conics take on the familiar shapes of ellipses, hyperbolas, and parabolas, as discussed in Section 3.5. Let $\underline{\mathbf{H}}$ (for *horizon*) be the preimage of $\mathbf{L}_\infty$. Let us assume for a moment that $\underline{\mathbf{H}}$ does not intersect the conic. Let $\underline{\mathbf{b}}[r,s]$ on $\underline{\mathbf{H}}$ be the pole to the polar $\underline{\mathbf{B}}[r,s]$. The affine images of $\underline{\mathbf{B}}[r,r]$ and $\underline{\mathbf{B}}[s,s]$ will be two *parallel tangents*, while $\underline{\mathbf{B}}[r,s]$ will be mapped to a secant of the affine conic. We call the tangents $\underline{\mathbf{B}}[r,r]$ and $\underline{\mathbf{B}}[s,s]$ *conjugate* to the secant $\underline{\mathbf{B}}[r,s]$. See Figure 4.11 for an illustration.

Affine conics have an important *shape measure*: we can determine how much a conic arc deviates from a circular arc. This deviation is known as the *eccentricity* ϵ of the conic. It is given by

$$\epsilon = 16\mathcal{A}^2 w^2 (1 - w^2), \tag{4.39}$$

where $\mathcal{A} = \text{area}(\mathbf{b}_0, \mathbf{b}_1, \mathbf{b}_2)$, quoted from [81]. The eccentricity is negative, zero, or positive, if the conic is a hyperbola, a parabola, or an ellipse, respectively. In particular, it equals one in case the conic is a circle.

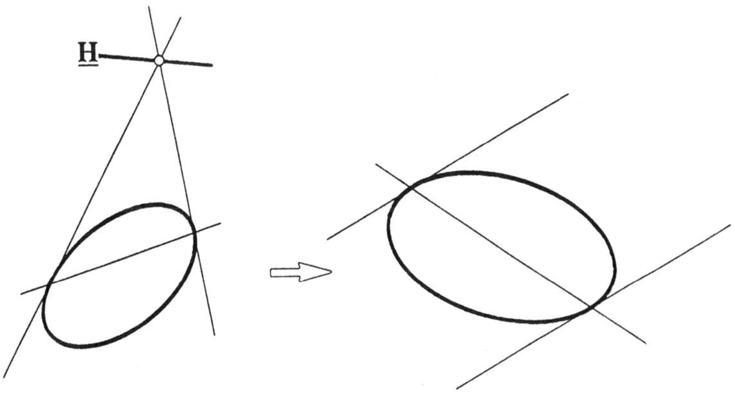

Figure 4.11. Conjugate tangents and secants: if a projective pole is mapped to $\mathbf{L}_\infty$, its corresponding tangents become parallel.

4.12 Problems

1. Show that the point $\underline{\mathbf{b}_0^2}$ defined by (4.9) does in fact lie on the tangent $\underline{\mathbf{T}}$ defined by (4.10), i.e., show that

$$\underline{\mathbf{T}\mathbf{b}_0^2} = 0.$$

2. Show that the implicit form of an affine conic is given by

$$\mathbf{x}^{\mathrm{T}} A \mathbf{x} + 2\mathbf{b}^{\mathrm{T}}\mathbf{x} + d = 0,$$

with $\mathbf{x} = [x, y]^{\mathrm{T}}$, A a 2×2 symmetric matrix, and $\mathbf{b} = [b, c]^{\mathrm{T}}$. Hint: use (4.6).

3. In Section 4.7, there was no requirement that the parameter values t_i be increasing. Investigate what will happen if, for the configuration of Figure 4.5, the t_i are given in nondecreasing order. Example: $0.2, 0.6, 0.4, 0.8$.

4. Use the concept of the complementary segment to prove the special case of Brianchon's theorem as shown in Figure 3.13.

5

Rational Quadratic Conics

While the projective treatment of conics reveals most of their fundamental properties in an elegant way, "real world" applications of conics happen in an affine environment. In this chapter, we will thus shift our emphasis from the projective treatment of conics to an affine viewpoint. [1]

5.1 The Rational Quadratic Form

In the projective plane $\mathbb{P}^2$, a conic may be described by a quadratic Bézier curve:

$$\begin{aligned} \underline{\mathbf{b}}^2(t) &= (1-t)^2\underline{\mathbf{b}}_0 + 2t(1-t)\underline{\mathbf{b}}_1 + t^2\underline{\mathbf{b}}_2 \\ &= (1-t)^2\begin{bmatrix} x_0 \\ y_0 \\ z_0 \end{bmatrix} + 2t(1-t)\begin{bmatrix} x_1 \\ y_1 \\ z_1 \end{bmatrix} + t^2\begin{bmatrix} x_2 \\ y_2 \\ z_2 \end{bmatrix} \\ &= (1-t)^2 z_0\begin{bmatrix} \mathbf{b}_0 \\ 1 \end{bmatrix} + 2t(1-t)z_1\begin{bmatrix} \mathbf{b}_1 \\ 1 \end{bmatrix} + t^2 z_2\begin{bmatrix} \mathbf{b}_2 \\ 1 \end{bmatrix}, \end{aligned}$$

[1] Here, we are dealing exclusively with planar curves. Hence we denote the homogeneous coordinate by z, as opposed to the homogeneous w in subsequent chapters.

where the $\mathbf{b}_i$ are already written as affine points:

$$\mathbf{b}_i = \frac{1}{z_i}\begin{bmatrix} x_i \\ y_i \end{bmatrix}. \tag{5.1}$$

We can now project our projective curve $\underline{\mathbf{b}}^2(t)$ into the (extended) affine plane, yielding the *rational quadratic form of a conic:*

$$\mathbf{b}^2(t) = \frac{(1-t)^2 z_0 \mathbf{b}_0 + 2t(1-t) z_1 \mathbf{b}_1 + t^2 z_2 \mathbf{b}_2}{(1-t)^2 z_0 + 2t(1-t) z_1 + t^2 z_2}. \tag{5.2}$$

The (affine) points $\mathbf{b}_i$ are called *control points*, and the z_i are called *weights*.

The projective shoulder tangent $\underline{\mathbf{T}}(\frac{1}{2})$ is mapped to the tangent $\mathbf{T}(\frac{1}{2})$ of the affine conic; the points $\underline{\mathbf{q}}_i$ are mapped to affine points $\mathbf{q}_i$:

$$\mathbf{q}_0 = \frac{z_0\mathbf{b}_0 + z_1\mathbf{b}_1}{z_0 + z_1}, \quad \mathbf{q}_1 = \frac{z_1\mathbf{b}_1 + z_2\mathbf{b}_2}{z_1 + z_2}. \tag{5.3}$$

Because of their dependence on the weights, we call the $\mathbf{q}_i$ *weight points*. The relation between weights and weight points is given by

$$\text{ratio}(\mathbf{b}_0, \mathbf{q}_0, \mathbf{b}_1) = \frac{z_1}{z_0} \quad \text{and} \quad \text{ratio}(\mathbf{b}_1, \mathbf{q}_1, \mathbf{b}_2) = \frac{z_2}{z_1}.$$

For the definition of ratios, see Section 2.7.

In the affine case, the *signs* of the weights are of considerable importance; we will require that they are all of the same sign. To see why this is desirable, let us investigate the case of z_0 being negative, while z_1 and z_2 are positive. Referring to Figure 5.1, we see that the denominator of (5.2), $z(t) = (1-t)^2 z_0 + 2t(1-t)z_1 + t^2 z_2$ has a zero in the interval $[0, 1]$. Assuming finite $\mathbf{b}_i$, this will cause the conic to have a singularity, i.e., it will cross the line at infinity for a value of $t^\star$ in $[0, 1]$. The tangent at $t^\star$ is an asymptote of the conic, which consequently is a hyperbola, as discussed in Section 3.5.

Unless otherwise stated, we will assume that all weights are positive,[2] even at the expense of leaving out hyperbolas with an asymptote in the interval $[0, 1]$. From Section 4.9, we know that it is then no restriction to assume that $z_0 = z_2 = 1$, and we will also assume that conics are given in this *standard form* from now on.

5.2 Conic Classification

An affine conic in standard form has a remarkable geometric property: its shoulder tangent is parallel to the chord $\overline{\mathbf{b}_0\mathbf{b}_2}$. This follows from the

[2]They may also be all negative — a common factor of -1 does not change anything!

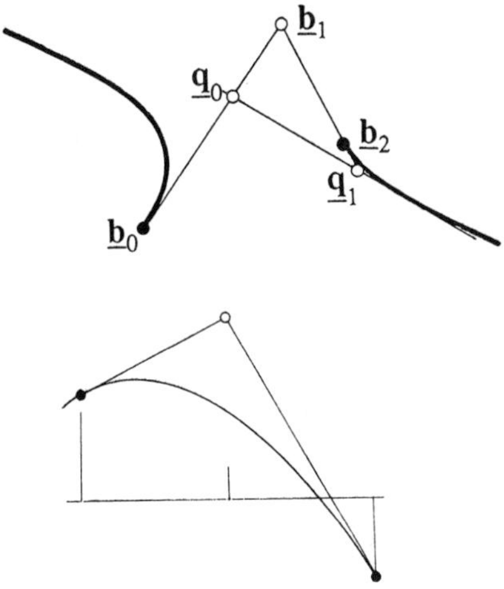

Figure 5.1. Weight signs: for $z_0 < 0$ and $z_1, z_2 > 0$, the conic will have a singularity in $[0, 1]$.

observation that

$$\text{ratio}(\mathbf{b}_0, \mathbf{q}_0, \mathbf{b}_1) = \text{ratio}(\mathbf{b}_2, \mathbf{q}_1, \mathbf{b}_1) = z.$$

Let us again denote by $\underline{\mathbf{H}}$ the projective line that corresponds to the affine line at infinity. Then we know that the projective lines $\underline{\mathbf{q}}_0 \wedge \underline{\mathbf{q}}_1$ and $\underline{\mathbf{b}}_0 \wedge \underline{\mathbf{b}}_2$ intersect on $\underline{\mathbf{H}}$, since that is the condition for them to be mapped to parallel lines in affine space. So if we standardize an affine conic by selecting a shoulder tangent parallel to $\overline{\mathbf{b}_0\mathbf{b}_2}$, this corresponds to choosing a projective shoulder tangent that passes through $[\underline{\mathbf{b}}_0 \wedge \underline{\mathbf{b}}_2] \wedge \underline{\mathbf{H}}$. Figure 5.2 shows three configurations, yielding the three affine conic types: the hyperbola for $\underline{\mathbf{H}}_h$, the parabola for $\underline{\mathbf{H}}_p$, the ellipse for $\underline{\mathbf{H}}_e$.

Referring to Figure 4.4, we can also say that a *parabola* in standard form goes to infinity for the value $t = \pm\infty$. The situation for a *hyperbola* in standard form is more involved; see Figure 5.3.

The weight function $z(t)$ provides us with a more algebraic means to classify rational quadratic conics into the three affine types: ellipses, hyperbolas, and parabolas. The singularities of the conic, i.e., the points where it meets the line at infinity, correspond to the zeroes of $z(t)$. These zeroes, if they exist, are outside the interval $[0, 1]$, and Figure 5.4 shows the three possibilities:

- either $z_1 < 1$, then there are no zeroes, and the conic is an ellipse,

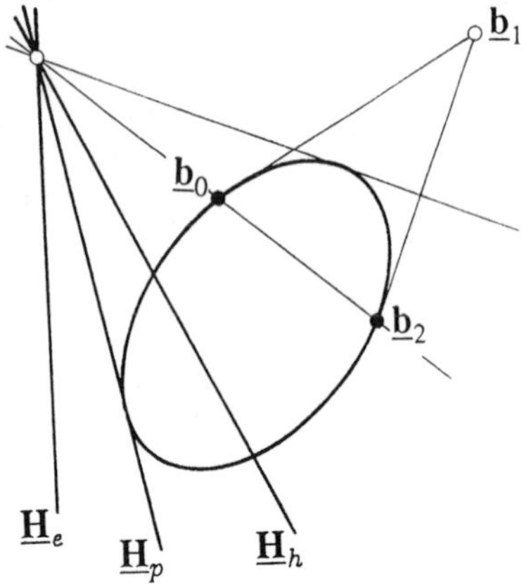

Figure 5.2. Conics in standard form: the projective preimage configuration of the three affine types.

- or $z_1 > 1$, then there are two finite zeroes, symmetric with respect to $t = 1/2$, and the conic is a hyperbola,
- or $z_1 = 1$, then there are two infinite zeroes, and the conic is a parabola.

The counting of the zeroes for the parabola weight function should be interpreted as the transition from the hyperbolic to the elliptic case: as z approaches 1 from above or from below, the two zeroes tend to infinity.

5.3 Properties of Rational Quadratics

In this section, we list some properties exhibited by all conics in rational quadratic form; properties of the different types are discussed at the end of this chapter. We assume that the conics are given in standard form.

Convex hull property: The conic segment corresponding to $t \in [0, 1]$ lies in the *convex hull* of the control polygon, i.e., it is contained inside the triangle formed by the control polygon. A stronger statement can be made also: the curve segment is contained inside the quadrilateral $\mathbf{b}_0, \mathbf{q}_0, \mathbf{q}_1, \mathbf{b}_2$, where the $\mathbf{q}_i$ are the weight points from (5.3).

Affine invariance: The following two procedures yield the same result. First, evaluate $\mathbf{b}(t)$ and map it to $\Phi\mathbf{b}(t)$ by an affine map (see Section 2.7). Second, map all $\mathbf{b}_i$ to $\Phi\mathbf{b}_i$ (leaving the weight z_1 unchanged)

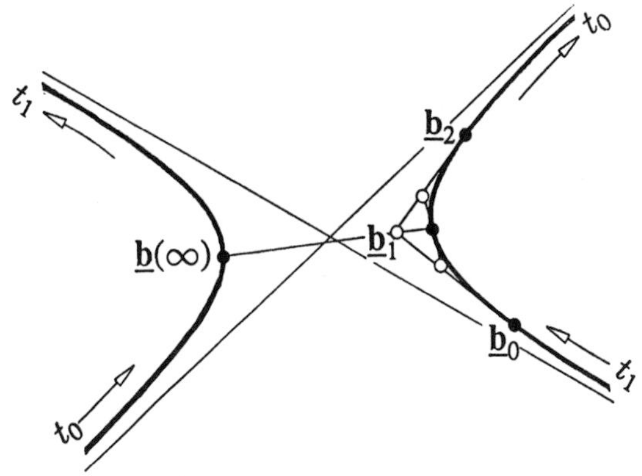

Figure 5.3. Hyperbolas in standard form: the figure shows how the parameter t traverses the curve.

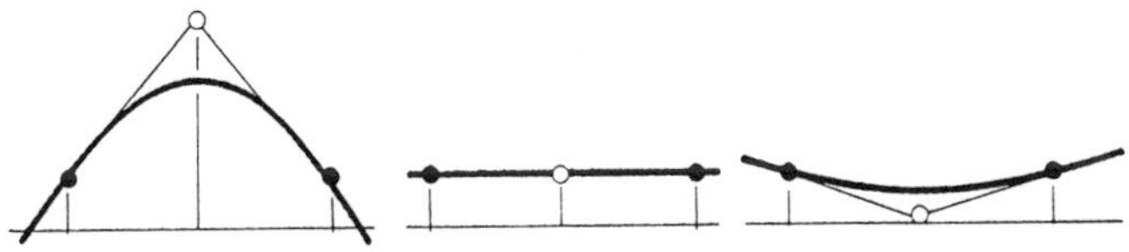

Figure 5.4. Conic classification: the zeroes of the weight function determine the type of conic. From left to right: weight functions for a hyperbola, a parabola, and an ellipse.

and evaluate the new rational quadratic at t. Note that the type of conic is not changed by an affine map.

Projective invariance: In the same sense, rational quadratics are projectively invariant. Note, however, that the ratios ratio$(\mathbf{b}_i, \mathbf{q}_i, \mathbf{b}_{i+1})$ and thus the weights will be changed after application of a projective map. Hence also the type (ellipse, etc.) of the conic may be changed.

Subdivision: The shoulder point $\mathbf{b}(\frac{1}{2})$ subdivides the curve into two pieces, a "left" and a "right" one. The control polygon for the left segment is given by $\mathbf{b}_0, \mathbf{q}_0 = \mathbf{b}_0^1(\frac{1}{2}), \mathbf{b}(\frac{1}{2})$, with weights $1, (1+z)/2, \frac{1}{2} + \frac{1}{4}z^2$. The right segment is treated analogously. Note that neither segment's weight is now in standard form.

5.4 Derivatives

Differentiation of conics is easy in projective space: there, they are quadratic polynomials, and their derivatives are linear; see Section 4.4. In affine space, however, we are dealing with rational quadratic polynomials, which would need the quotient rule for taking derivatives. A simple trick avoids this unpleasant task: our conic $\mathbf{c}(t)$ is of the form $\mathbf{c}(t) = \mathbf{p}(t)/z(t)$, with a polynomial numerator $\mathbf{p}(t)$. Thus

$$\mathbf{p}(t) = z(t)\mathbf{c}(t).$$

This polynomial curve is differentiated using the product rule:

$$\dot{\mathbf{p}}(t) = \dot{z}(t)\mathbf{c}(t) + z(t)\dot{\mathbf{c}}(t),$$

the dot denoting differentiation with respect to t. The expression $\dot{\mathbf{c}}(t)$ on the right hand side is our desired conic derivative! We can solve for it and obtain

$$\dot{\mathbf{c}}(t) = \frac{1}{z(t)}[\dot{\mathbf{p}}(t) - \dot{z}(t)\mathbf{c}(t)]. \tag{5.4}$$

This procedure may be repeated to yield higher order derivatives; see Section 7.8 for details.

We may evaluate (5.4) at the endpoints $t = 0$ and $t = 1$:

$$\dot{\mathbf{c}}(0) = \frac{2z_1}{z_0}\Delta\mathbf{b}_0, \quad \dot{\mathbf{c}}(1) = \frac{2z_1}{z_2}\Delta\mathbf{b}_1. \tag{5.5}$$

5.5 Control Vectors

It is possible that one or two of the control points $\underline{\mathbf{b}}_i$ of the projective conic are on $\underline{\mathbf{H}}$, the preimage of the affine line at infinity. They are then mapped to *control vectors.* Figure 5.5 gives an example. Varying the length of a control vector has a predictable effect on the conic, as illustrated in Figure 5.6.

Suppose that $\mathbf{b}_1$ is a control vector while $\mathbf{b}_0$ and $\mathbf{b}_2$ are control points; thus $\underline{\mathbf{b}}_1$'s third component z_1 is zero. We can then rewrite the rational form of this conic:

$$\mathbf{b}(t) = \frac{z_0\mathbf{b}_0 B_0^2(t) + z_2\mathbf{b}_2 B_2^2(t)}{z_0 B_0^2(t) + z_2 B_2^2(t)} + \frac{\mathbf{b}_1 B_1^2(t)}{z_0 B_0^2(t) + z_2 B_2^2(t)}.$$

For a more general treatment, see Section 7.10

Control vectors were first considered by K. Vesprille [115], and later by L. Piegl [89]. Both authors use the term "infinite control points." As we

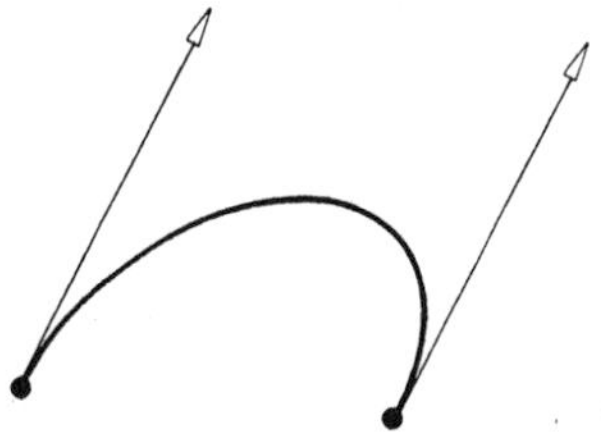

Figure 5.5. Control vectors: an ellipse with a control vector $\mathbf{b}_1$.

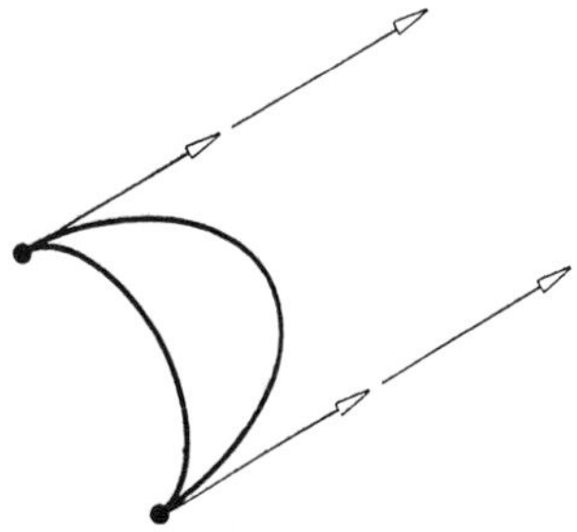

Figure 5.6. Control vectors: the effect of varying $\mathbf{b}_1$'s length.

have seen, an infinite point (short for a point at infinity) corresponds to a whole family of parallel vectors. Each vector in this family gives rise to a different curve, but corresponds to the same point at infinity. The "infinite control point" terminology is thus oversimplifying.[3] Fiorot and Jeannin [52] share this criticism.

If the representation of a conic involves control vectors, the convex hull property is lost – but the use of the weight points $\mathbf{q}_i$ recovers it: the curve segment is still inside the quadrilateral $\mathbf{b}_0, \mathbf{q}_0, \mathbf{q}_1, \mathbf{b}_2$.

It is also possible that $\underline{\mathbf{b}}_0$ and $\underline{\mathbf{b}}_2$ are on $\underline{\mathbf{H}}$. Then, the resulting affine conic crosses the line at infinity twice, and thus is a hyperbola. Figure 5.7 illustrates.

Finally, a practical note on control vectors: the current standard format for geometric definition in CAD/CAM systems, IGES,[4] does not allow zero weights; see also Section 15.1. This means that data exchange will be impossible if IGES is chosen as the neutral data format and control vectors

[3] T. DeRose brought this point to my attention.

[4] IGES is short for Initial Graphics Exchange Standard.

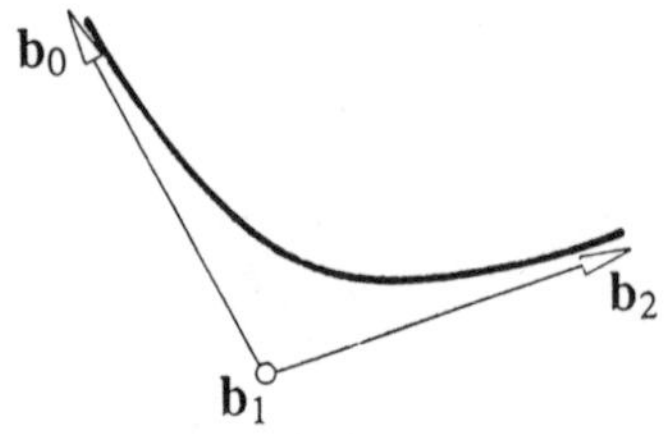

Figure 5.7. Control vectors: a hyperbola with two control vectors $\mathbf{b}_0, \mathbf{b}_2$ and a control point $\mathbf{b}_1$.

are involved. Some systems try to circumvent this by replacing the zero weights by very small constants, like 10^{-10}.

5.6 The Parabola

The parabola enjoys a special role among conics in rational quadratic form: it allows a polynomial, or integral form. In that form, its weight points are simply the midpoints of the control polygon legs. The de Casteljau algorithm simplifies since we can use ratios instead of cross ratios:

$$\operatorname{cr}(\mathbf{b}_i^r, \mathbf{b}_i^{r+1}, \mathbf{q}_i^r, \mathbf{b}_{i+1}^r) = \operatorname{ratio}(\mathbf{b}_i^r, \mathbf{b}_i^{r+1}, \mathbf{b}_{i+1}^r) = \frac{t}{1-t},$$

and the de Casteljau algorithm becomes

$$\mathbf{b}_i^{r+1}(t) = (1-t)\mathbf{b}_i^r(t) + t\mathbf{b}_{i+1}^r(t). \tag{5.6}$$

But we can also write a parabola as a true rational quadratic: employing the technique of *reparametrization* from Section 4.9, we can write it as

$$\mathbf{b}(t) = \frac{\mathbf{b}_0 B_0^2(t) + c\mathbf{b}_1 B_1^2(t) + c^2\mathbf{b}_2 B_2^2(t)}{B_0^2(t) + cB_1^2(t) + c^2 B_2^2(t)} \tag{5.7}$$

with an arbitrary nonzero constant c. Only for $c = 1$, do we obtain the polynomial form.

5.7 The Circle

Of all conics, the circular arc is the most widely used. Its control polygon must satisfy a special condition: it has to form an isosceles triangle, due

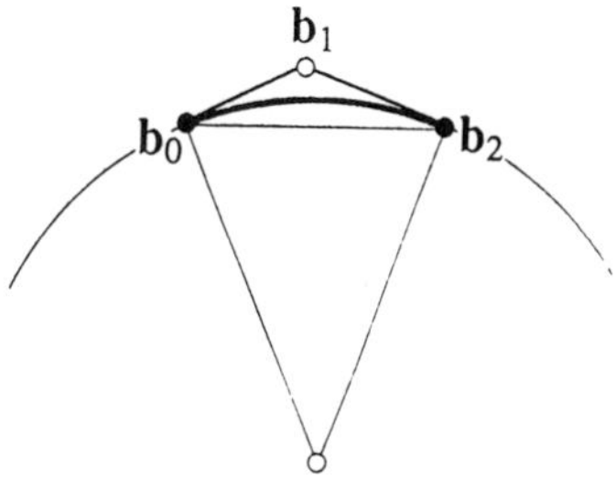

Figure 5.8. The circle: the geometry of the control polygon.

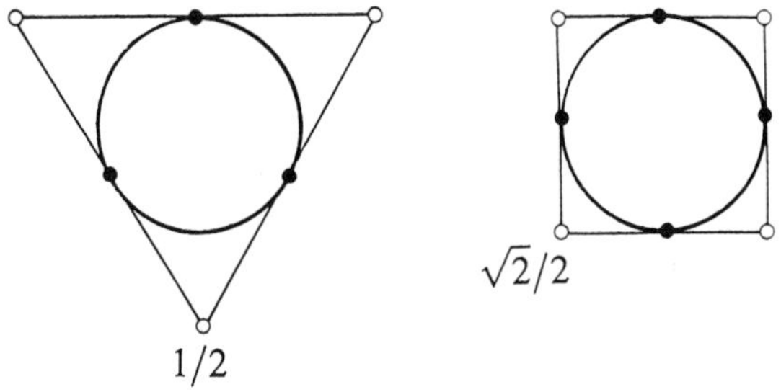

Figure 5.9. The full circle: it may be represented by three or four rational quadratics, with the weights of the control points indicated.

to the circle's symmetry properties. Referring to Figure 5.8, and assuming standard form, recall that $z_1 = \text{ratio}(\mathbf{b}_0, \mathbf{q}_0, \mathbf{b}_1)$. Thus

$$z_1 = \cos\alpha,$$

with $\alpha = \angle(\mathbf{b}_2, \mathbf{b}_0, \mathbf{b}_1)$.

In the case $\alpha = 90^o$, the weight vanishes and we have the control vector form of a conic, as outlined in Section 5.5.

A whole circle may be represented in many ways by piecewise rational quadratics. One example is to use the above form to write the "top" part, and then use the complementary segment from Section 4.10 to write the "bottom" part. It is probably more convenient — retaining the convex hull property for positive weights — to dissect the full circle into three or four parts, as shown in Figure 5.9.

Although we can write an arc of a circle in rational quadratic form, one should not overlook that we then sacrifice one nice property of the familiar sin/cos parametrization: in the rational quadratic form, the parameter t does not traverse the circle with *unit speed.* If a curve is traversed with unit speed, its parameter is the *arc length* parameter. Unit speed traversal of a curve means that the tangent vector $\dot{\mathbf{x}}(t)$ is of unit length.[5] Clearly, the derivative of a rational quadratic does not have this property. Thus, if an arc of a rational quadratic is to be split into a certain number of segments, each subtending the same angle, numerical techniques must be invoked. In the sin/cos parametrization, by contrast, equal parameter increments ensure equal subtended angles.

It can be shown that higher degree rational curves also do not have a constant length first derivative vector; see [51].

Circular arcs may be pieced together to form *circle splines*, with easy control of tangent continuity. These splines play a role in numerically controlled (NC) manufacturing, as some NC machines can handle circles in hardware. We list some of the literature on circle splines: [27], [85], [101], [109].

5.8 Circles and the Stereographic Projection

A well-known parametrization of the circle may be obtained as follows. Consider a unit circle with its center at the origin of a Euclidean coordinate system. We can write the southeast quarter as a rational quadratic with control points

$$\begin{bmatrix} 0 \\ -1 \end{bmatrix}, \begin{bmatrix} 1 \\ -1 \end{bmatrix}, \begin{bmatrix} 1 \\ 0 \end{bmatrix}$$

and standard weights $1, \frac{1}{2}\sqrt{2}, 1$. We may also reparametrize this circle representation, using a reparametrization constant c.[6] The equation then becomes:

$$\begin{bmatrix} x \\ y \end{bmatrix} = \begin{bmatrix} \frac{c\sqrt{2}(1-t)t+c^2t^2}{(1-t)^2+c\sqrt{2}(1-t)t+c^2t^2} \\ \frac{-(1-t)^2-c\sqrt{2}(1-t)t}{(1-t)^2+c\sqrt{2}(1-t)t+c^2t^2} \end{bmatrix},$$

with $c = 1$ for the familiar standard form. Let us try to arrange the parametrization such that the curve passes through $\mathbf{n} = [0, 1]^{\mathrm{T}}$, the north pole of the circle, for $t = \infty$. This leads to

$$\begin{bmatrix} 0 \\ 1 \end{bmatrix} = \begin{bmatrix} \frac{c^2-c\sqrt{2}}{c^2-c\sqrt{2}+1} \\ \frac{c\sqrt{2}-1}{c^2-c\sqrt{2}+1} \end{bmatrix},$$

---

[5] It will be of constant length if we are dealing with circles other than the unit circle.

[6] We are referring to a rational linear reparametrization as outlined in Section 4.9.

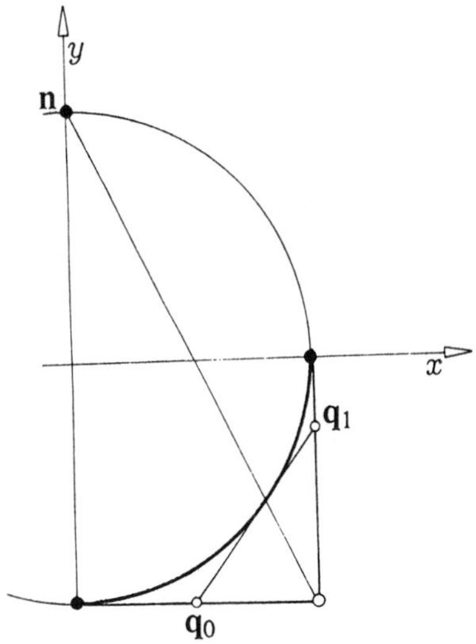

Figure 5.10. The circle: a parametrization that corresponds to the stereographic projection.

from which we deduce $c = \sqrt{2}$. We may then take the new Bézier form and rewrite it as follows:

$$\begin{bmatrix} x \\ y \end{bmatrix} = \begin{bmatrix} \frac{2t}{t^2+1} \\ \frac{t^2-1}{t^2+1} \end{bmatrix}. \tag{5.8}$$

This parametrization is an example of the *stereographic projection*: one easily verifies that the line through the north pole **n** and $[x(t), y(t)]^{\mathrm{T}}$ intersects the $x-$axis in $x = t$. Thus every point of the $x-$axis has a unique image point on the circle and vice versa: they are related by a stereographic projection through **n**. Figure 5.10 illustrates. More on these maps in Section 13.3.

5.9 Problems

1. In Figure 5.7, define a coordinate system with origin $\mathbf{b}_1$ and axis directions $\mathbf{b}_0$ and $\mathbf{b}_2$. Show that in this coordinate system, the hyperbola takes on the familiar form $y = 1/x$.

2. In Figure 5.2, $\underline{\mathbf{H}}$ could coincide with $\underline{\mathbf{q}}_0 \wedge \underline{\mathbf{q}}_1$. Illustrate the resulting affine configuration.

3. Find the reparametrization $\phi = \phi(t)$ of (5.8) such that we obtain the circle equation $[x(\phi), y(\phi)]^{\mathrm{T}} = [\cos(\phi), \sin(\phi)]^{\mathrm{T}}$.

6

Conic Splines

Rational quadratics may be "pieced together," so as to form composite curves that can be much more complex than single conics. Such curves are called *conic splines.* Where two pieces, or *segments*, of a conic spline meet, the curve will be of a certain smoothness: it could be analytically differentiable: C^1, C^2, or it could be geometrically smooth: G^1, G^2. The different cases are discussed in this chapter.

6.1 C^1 Conditions

Conics may be pieced together to form a more complicated curve, whose shape is too complex to be captured by a single conic. Such composite curves are called *conic splines.* When projected into affine space, it will be an instance of NURB curves.

Figure 6.1 shows two conics: one with control polygon $\underline{\mathbf{b}}_0, \underline{\mathbf{b}}_1, \underline{\mathbf{b}}_2$ and one with control polygon $\underline{\mathbf{b}}_2, \underline{\mathbf{b}}_3, \underline{\mathbf{b}}_4$. At $\underline{\mathbf{b}}_2$, they share a common tangent, i.e. $\det[\underline{\mathbf{b}}_1, \underline{\mathbf{b}}_2, \underline{\mathbf{b}}_3] = 0$.

Does this mean that they are C^1? In other words, do both conics share the same first derivative at $\underline{\mathbf{b}}_2$? Recall that the tangent to a curve is a line, whereas the derivative is a point, so it is not obvious that tangent continuity implies derivative continuity. Derivative continuity is a property

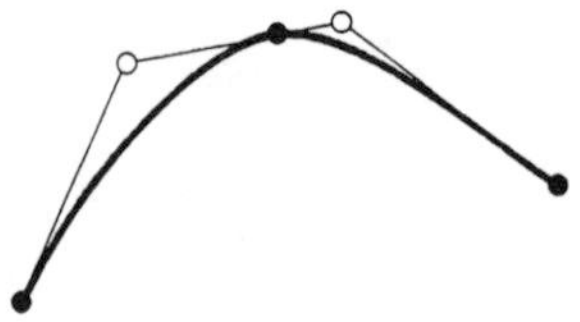

Figure 6.1. Piecewise conics: two conics form a smooth composite curve.

of the *parametrization* of a curve, i.e., more than just a geometric property. In order to formalize this, let each segment of our composite curve be defined over an interval[1] of $\mathbb{P}^1$: let the "left" conic be defined over $[u_0, u_1]$ and the "right" conic over $[u_1, u_2]$. Let us assume a point m_i in each of the two intervals such that they are equivalent reference frames; see Section 1.4. This means that the m_i have to satisfy

$$\mathrm{cr}(u_0, m_0, u_1, \infty) = \mathrm{cr}(u_1, m_1, u_2, \infty),$$

or

$$\frac{m_0 - u_0}{u_1 - u_0} = \frac{m_1 - u_1}{u_2 - u_1}.$$

We now say the two segments are C^1 if the two triples $\underline{\mathbf{b}}_1, \underline{\mathbf{q}}_1, \underline{\mathbf{b}}_2$ and $\underline{\mathbf{b}}_2, \underline{\mathbf{q}}_2, \underline{\mathbf{b}}_3$ form two equivalent reference frames for the curve's tangent at $\underline{\mathbf{b}}_2$ and if these frames are a projective map of the two equivalent domain frames u_0, m_0, u_1 and u_1, m_1, u_2. We thus obtain

$$\mathrm{cr}(u_0, m_0, u_1, u_2) = \mathrm{cr}(\underline{\mathbf{b}}_1, \underline{\mathbf{q}}_1, \underline{\mathbf{b}}_2, \underline{\mathbf{b}}_3) \tag{6.1}$$

and

$$\mathrm{cr}(u_0, u_1, m_1, u_2) = \mathrm{cr}(\underline{\mathbf{b}}_1, \underline{\mathbf{b}}_2, \underline{\mathbf{q}}_2, \underline{\mathbf{b}}_3) \tag{6.2}$$

as the two equations that have to be met for the left and right segment to be C^1 with respect to the *knot sequence* u_0, u_1, u_2.

Since the first derivative of a conic may be defined in terms of a blossom value at infinity (see Section 4.8), these conditions are equivalent to

$$\underline{\mathbf{b}}_{\text{left}}[u_1, \infty] \mathrel{\hat{=}} \underline{\mathbf{b}}_{\text{right}}[u_1, \infty]. \tag{6.3}$$

This interpretation of C^1 continuity is illustrated in Figure 6.2.

[1] We assume that $\mathbb{P}^1$ is endowed with a projective reference frame, such that each point on it corresponds to a real number, or parameter. An interval $[a, b]$ of $\mathbb{P}^1$ is then the set of all points whose parameter u satisfies $a \le u \le b$. We will also refer to u as a point in $\mathbb{P}^1$. Note that this interval, when projected into affine space, may contain the point at infinity.

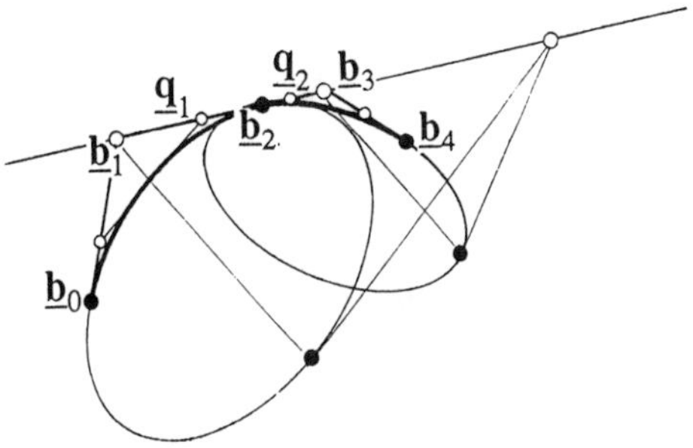

Figure 6.2. C^1 continuity: two conics form a differentiable piecewise curve at $\underline{\mathbf{b}}_2$ if the corresponding tangent has the same intersection with both conics' tangents at infinity.

Next, we might want to ask about C^2 continuity of our composite curve.[2] The algebraic condition is that the second derivatives of both curves match:

$$\underline{\mathbf{b}}_{\text{left}}[\infty, \infty] \mathrel{\hat{=}} \underline{\mathbf{b}}_{\text{right}}[\infty, \infty]. \tag{6.4}$$

Thus our conics, or quadratic curves, agree in position, first and second derivative at u_1. It follows that both conics have to be part of *one* conic.

An interesting consequence is that C^1 continuity between two conics implies C^2 continuity provided that they are both part of the same conic.

Figure 6.3 shows a conic that is covered by three C^2 quadratics. Note how each conic has the same point $\underline{\mathbf{i}}$ corresponding to $u = \infty$. In that figure, the points $\underline{\mathbf{q}}_2$ and $\underline{\mathbf{q}}_3$ are "outside" the polygon legs $\underline{\mathbf{b}}_2, \underline{\mathbf{b}}_3$ and $\underline{\mathbf{b}}_3, \underline{\mathbf{b}}_4$, respectively. When projecting into affine space, this will result in negative weights.

6.2 Parametric Continuity for Rational Conic Splines

In affine space, conics are represented by rational quadratics, and conic splines are piecewise rational quadratics. The above C^1 and C^2 conditions still hold, as derivative continuity is preserved by projecting from projective into affine space. But there are more general continuity conditions for rational splines: we may use the *quotient rule* for differentiating each segment! Assuming the "left" conic segment, defined over the interval $[u_0, u_1]$, has the control polygon $\mathbf{b}_0, \mathbf{b}_1, \mathbf{b}_2$, and the "right" segment, defined over

[2]When we talk about C^2 continuity, we always assume that the curve is already C^1 and C^0.

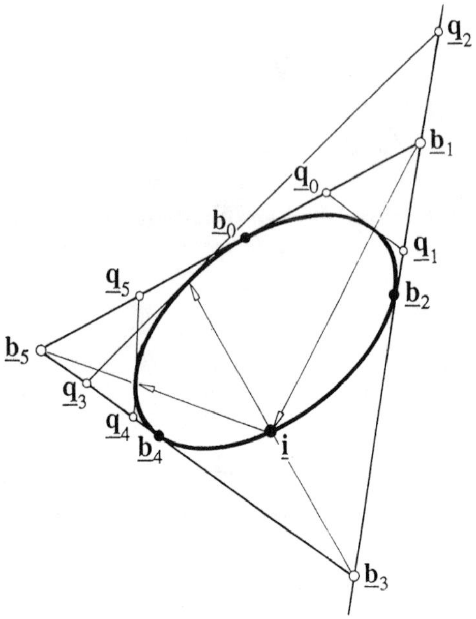

Figure 6.3. C^2 conics: a conic is covered by three quadratic segments.

$[u_1, u_2]$, is given by $\mathbf{b}_1, \mathbf{b}_2, \mathbf{b}_3$. Also assume that each control point $\mathbf{b}_i$ has a weight z_i attached to it.

The two rational quadratic conics form a C^1 curve over the domain $[u_0, u_2]$ if

$$\frac{z_1 \Delta \mathbf{b}_1}{\Delta_0} = \frac{z_3 \Delta \mathbf{b}_2}{\Delta_1}, \tag{6.5}$$

where we have set $\Delta_i = u_{i+1} - u_i$. This condition is more general than the one obtained by demanding that both segments form a C^1 curve in projective space: that would result in C^1 continuity of each component function $x(u), y(u), z(u)$.

6.3 Conic B-splines

When two conics meet such that they satisfy the strict C^1 condition (6.3) we know that each control point $\mathbf{b}_{2i}$ is determined from (6.3), as is each weight z_{2i}:

$$\underline{\mathbf{b}}_{2i} = \frac{\Delta_i \underline{\mathbf{b}}_{2i-1} + \Delta_{i-1} \underline{\mathbf{b}}_{2i+1}}{\Delta_{i-1} + \Delta_i}. \tag{6.6}$$

In order to *store* this spline, it thus seems sufficient to store only the $\underline{\mathbf{b}}_{2i\pm 1}$ instead of all control points. In doing so, we will relabel them as $\underline{\mathbf{d}}_i$; see Figure 6.4. We now also make the transition to *rational splines*, as this is closer to the widely accepted IGES format. Thus instead of dealing with projective points $\underline{\mathbf{d}}_i$, we are dealing with affine points $\mathbf{d}_i$. These are obtained by dividing the first two coordinates of each $\underline{\mathbf{d}}_i$ by its third coordinate v_i. The v_i are then stored separately and are called *weights*.

A rational conic spline could be defined as follows:

1. the number N of conic segments,
2. by a sequence of $N+2$ control points $\mathbf{d}_0, \ldots, \mathbf{d}_{N+1}$,
3. by a sequence of weights $v_0, \ldots, v_{N+1}$,
4. by a sequence of knots $u_0, \ldots, u_N$.

The last item, the *knots*, needs some explanation. We assume that the ith conic segment is defined over the interval $[u_i, u_{i+1}]$. The increasing sequence $\{u_i\}$ of reals is called the *knot sequence.* It forms a partition of the domain $[u_0, u_N]$ of the conic spline $\mathbf{s}(u)$.

The term B-spline will be explained shortly; for now, note that we have reduced the amount of storage compared to the previous data specification.

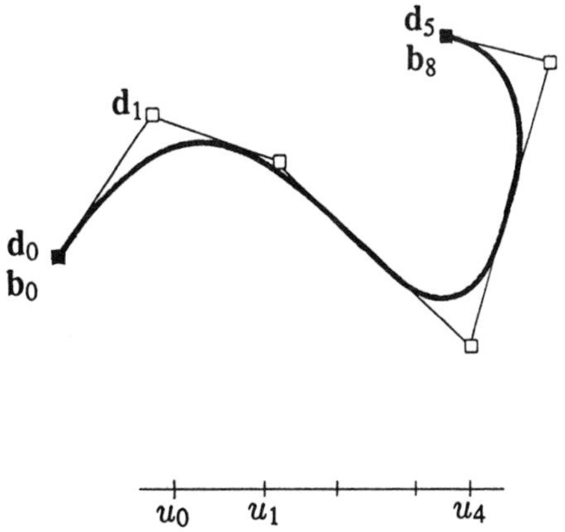

Figure 6.4. Conic B-splines: control polygon, curve, and knot sequence.

Also, we are guaranteed to represent curves that adhere to the strict C^1 condition (6.3), as shown in Figure 6.4. This data specification is essentially what we will encounter in Section 15.1 as the IGES format.

Note that conic splines that merely satisfy the weaker C^1 condition (6.5) do not allow such a compact representation: the weights z_{2i} are not tied to their neighboring weights by any condition – only the control points $\mathbf{b}_{2i}$ are![3]

Now to the term "B-spline," which is short for "basis spline." It is justified by the fact that we may write a strictly C^1 conic spline curve in this form:

$$\mathbf{s}(u) = \frac{\sum_{i=0}^{N+2} v_i \mathbf{d}_i N_i^2(u)}{\sum_{i=0}^{N+2} v_i N_i^2(u)}. \tag{6.7}$$

The functions N_i^2 are the quadratic B-splines that will be introduced in Chapter 10.

6.4 Curvature Continuity

The most significant geometric feature of any planar curve in Euclidean space is its *curvature*, or the rate of change of its tangent direction. We refer the reader to [25], [40], or [47] for the necessary differential geometry. The curvature $\kappa(t)$ of a parametric curve $\mathbf{x}(t)$ in $\mathbb{E}^2$ is given by

$$\kappa(t) = \frac{||\dot{\mathbf{x}} \wedge \ddot{\mathbf{x}}||}{||\dot{\mathbf{x}}||^3}. \tag{6.8}$$

The curvature is the inverse of the radius of the *osculating circle*, which is the circle that best approximates a curve at a given point. Where the curve is bent sharply, its curvature is high; where the curve is flat, its curvature is low. Note that curvature is a Euclidean concept; it cannot be defined in a straightforward manner in projective space. The curvature of a curve changes under affine and projective transformations; only (Euclidean) rotations and translations — so-called rigid body motions — leave it unchanged.[4]

Let us now try to evaluate a conic's curvature at the endpoints of its control polygon.[5] Inserting the appropriate quantities for the derivatives of a rational quadratic into (6.8), and evaluating at $t = 0$ and $t = 1$, we

[3] In terms of the IGES format from Section 15.1, this means that all u_i are treated as *double knots*.

[4] But curvature *continuity* is invariant under all these transformations!

[5] When talking about curvature, we always refer to conics in Euclidean space – projective concepts are of little use here!

obtain

$$\kappa(0) = \frac{\mathcal{A}}{z_1^2 \ell_0^3}, \quad \kappa(1) = \frac{\mathcal{A}}{z_1^2 \ell_1^3}, \tag{6.9}$$

where $\mathcal{A}$ denotes the area of the triangle formed by the conic control polygon, and ℓ_0, ℓ_1 are the polygon leg lengths, as shown in Figure 6.5. We assume that the conic is in standard form. Note that we can assign a *sign* to the curvature, depending on the sign of $\mathcal{A}$. See Figure 6.5 for an illustration. The sign will not change on ellipses or parabolas, but the two branches of a hyperbola will have curvatures of opposite signs.

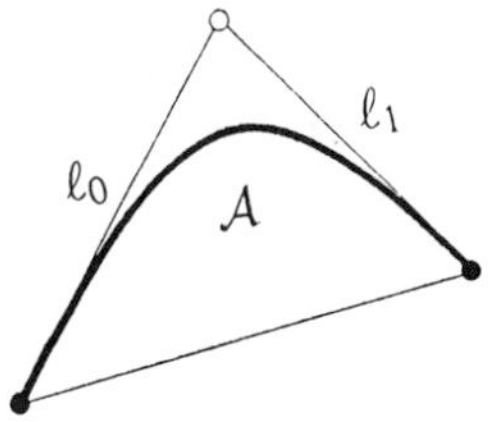

Figure 6.5. Conic curvatures: the curvature of a conic at its endpoints is determined by the shown geometric elements.

In addition to the endpoints, it is also easy to compute the curvature $\kappa(\frac{1}{2})$ of the shoulder point $\mathbf{b}(\frac{1}{2})$. It is given by

$$\kappa(\frac{1}{2}) = 8z_1 \frac{\mathcal{A}}{||\mathbf{b}_2 - \mathbf{b}_0||^3}. \tag{6.10}$$

6.5 G^2 Conic Splines

Two coplanar conics in standard form, with control polygons $\mathbf{b}_0, \mathbf{b}_1, \mathbf{b}_2$ and $\mathbf{b}_2, \mathbf{b}_3, \mathbf{b}_4$ and center weights z_1, z_3 are *curvature continuous* if

$$\frac{\mathcal{A}_1}{z_1^2 \ell_1^3} = \frac{\mathcal{A}_3}{z_3^2 \ell_2^3}, \tag{6.11}$$

where $\mathcal{A}_i = \text{area}(\mathbf{b}_{i-1}, \mathbf{b}_i, \mathbf{b}_{i+1})$ and $\ell_i = \text{length}(\mathbf{b}_i, \mathbf{b}_{i+1})$.

Note that (6.11) implies that the composite curve must be *convex*, since the signs of $\mathcal{A}_1$ and of $\mathcal{A}_3$ must be the same! We also use the term G^2 continuity, or second order geometric continuity to describe this type of smoothness.

If we multiply the central weights z_1 and z_3 of our composite conic by the same factor, we will maintain curvature continuity. Thus, we may thus

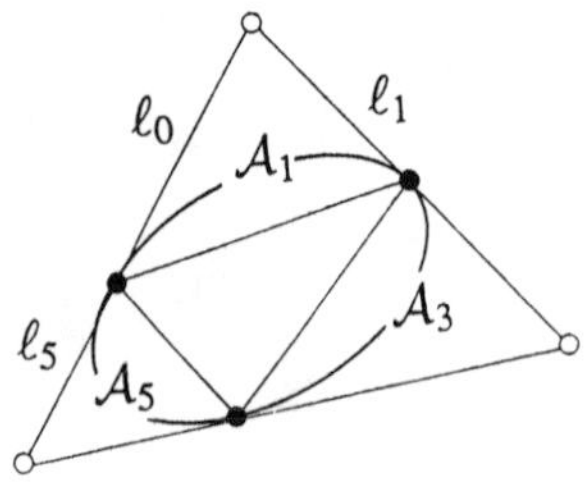

Figure 6.6. Closed conic splines: curvature continuity imposes additional conditions on the curve's geometry.

pull the piecewise curve closer to or push it away from the control polygon, still maintaining curvature continuity.

An interesting feature[6] of conics is that the *ratio* of $\kappa(0)$ and $\kappa(1)$ is independent of the weight z_1:

$$\frac{\kappa(0)}{\kappa(1)} = \left[\frac{\ell_1}{\ell_0}\right]^3. \tag{6.12}$$

This has an implication on *closed* conic splines: consider the closed curve shown in Figure 6.6. Curvature continuity demands that the following three conditions be met:

$$\begin{aligned}
\frac{\mathcal{A}_1}{z_1^2\ell_1^3} &= \frac{\mathcal{A}_3}{z_3^2\ell_2^3},\\
\frac{\mathcal{A}_3}{z_3^2\ell_3^3} &= \frac{\mathcal{A}_5}{z_5^2\ell_4^3},\\
\frac{\mathcal{A}_5}{z_5^2\ell_5^3} &= \frac{\mathcal{A}_1}{z_1^2\ell_0^3}.
\end{aligned}$$

If we multiply all left-hand sides and all right-hand sides, we obtain:

$$\ell_1\ell_3\ell_5 = \ell_2\ell_4\ell_0.$$

This condition — with an obvious extension for curves with more than three segments — limits the geometry of closed conic spline curves with curvature continuity. If only the $\mathbf{b}_{2i+1}$ were given, and we were asked for permissible locations for the $\mathbf{b}_{2i}$, one possible choice for them is to set

$$\frac{\ell_{2i-1}}{\ell_{2i}} = \frac{\mathcal{A}_{2i-1}}{\mathcal{A}_{2i+1}},$$

where $\mathcal{A}_{2i+1}$ is the area of the triangle formed by $\mathbf{b}_{2i}, \mathbf{b}_{2i+1}, \mathbf{b}_{2i+2}$.

[6] This property was pointed out to me by a referee who suggested improvements on the paper [45].

6.6 Interpolation with G^2 Conic Splines

We now consider an interpolation problem where the given data are:

- two points $\mathbf{b}_0$ and $\mathbf{b}_4$,
- a tangent at $\mathbf{b}_0$ and $\mathbf{b}_4$,
- the intersection $\mathbf{d}$ of the two tangents,
- curvatures κ_0 and κ_2 at $\mathbf{b}_0$ and $\mathbf{b}_4$,
- another tangent $\mathbf{T}$.

This configuration is assumed to be convex. Can we find a conic that interpolates to this data set? In general, we cannot — the given information is too much for one conic to cope.

But we may use *two* conics: this approach is due to H. Pottmann [93]. He constructs two G^2 conics in standard form that join at a point $\mathbf{b}_2$; see Figure 6.7.

With the notation from that figure, the condition for curvature continuity becomes

$$\frac{\mathcal{A}_1}{z_1^2 \ell_1^3} = \frac{\mathcal{A}_3}{z_3^2 \ell_2^3},$$

from which we conclude that

$$\frac{\ell_0^3}{\ell_3^3} \frac{\kappa_0}{\kappa_2} = \frac{\ell_1^3}{\ell_2^3}.$$

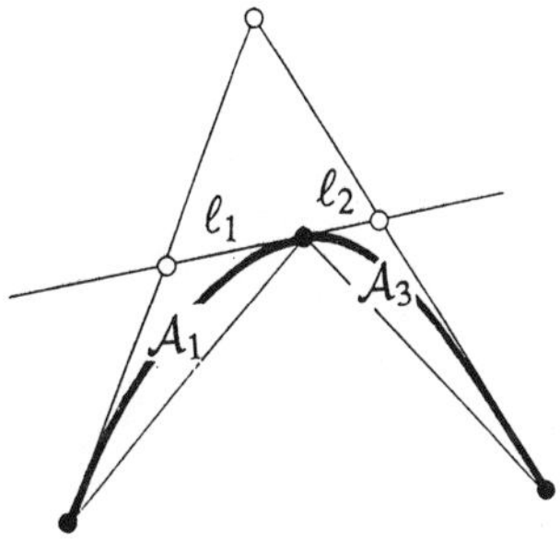

Figure 6.7. Conic interpolants: two conics may be used to interpolate to given positions, tangents, and curvatures.

This allows us to compute ℓ_1/ℓ_2, and thus we have found the point $\mathbf{b}_2$. The missing weights z_1 and z_3 are found from the interpolatory conditions

$$\kappa_0 = \frac{\mathcal{A}_1}{z_1^2 \ell_0^2} \quad \text{and} \quad \kappa_2 = \frac{\mathcal{A}_3}{z_3^2 \ell_3^2}.$$

The tangent $\mathbf{T}$ might not be given in a particular application, and Pottmann [93] points out several ways to automate its choice.

When is it possible to solve our initial problem with *one* conic? For that to be possible, the prescribed data cannot be independent; rather, they must satisfy the restriction

$$\frac{\kappa_0}{\kappa_2} = \frac{||\mathbf{d} - \mathbf{b}_0||^3}{||\mathbf{d} - \mathbf{b}_4||^3},$$

as follows from (6.12). The number

$$I = \frac{\kappa_2}{\kappa_0} \frac{||\mathbf{d} - \mathbf{b}_0||^3}{||\mathbf{d} - \mathbf{b}_4||^3} \tag{6.13}$$

thus measures the "nonconicness" of the given data. If it equals one, they permit an interpolating conic.

Pottmann [93] made the interesting geometric observation that the number I is invariant under projective transformations. In particular, the property $I = 1$ (indicating curvature continuity)is a projective invariant, a fact

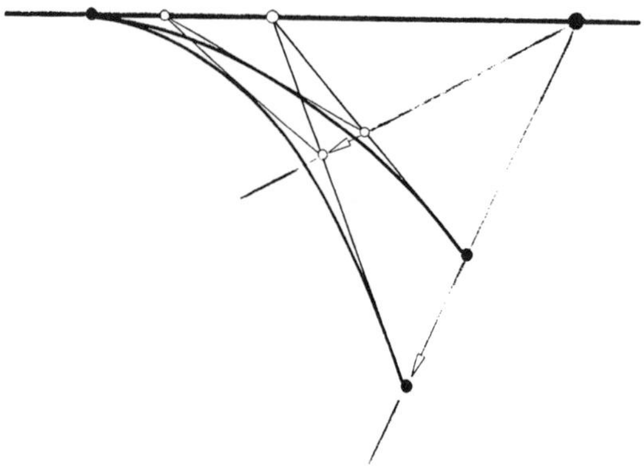

Figure 6.8. G^2 continuity: the two shown conics are related by a central collineation.

that is known as the Memke-Smith theorem. This theorem is somewhat surprising since individual curvatures are certainly not projectively invariant as they involve length and area computations.

6.7 Higher Order Smoothness

We have seen how to join conics so as to achieve G^1 or G^2 continuity. There are more ways of doing this; we report an approach that is due to Geise and Juettler [60].

Referring to Figure 6.8, we see two conics (in projective space) that share a common tangent $\underline{\mathbf{T}}$. Pick a point $\underline{\mathbf{c}}$ on this tangent, and define a *central collineation* with axis $\underline{\mathbf{T}}$ and center $\underline{\mathbf{c}}$. Then the two conics shown are G^2 continuous in the following sense: In a "small" neighborhood of $\underline{\mathbf{b}}_0$, pick three points on each conic. If the neighborhood is chosen small enough, the three points will coincide. This is called "contact of order two" and is the most general definition of G^2 continuity. In the context of affine parabolas, a commonly used G^2 condition is to choose $\mathbf{c}$ at infinity; then the central collineation becomes a shear.

An increase in the smoothness between two conics may be achieved as follows: referring to Figure 6.9, we now choose a central collineation with center $\underline{\mathbf{b}}_0$. This ensures that the conics have G^3 continuity, or contact of order three.

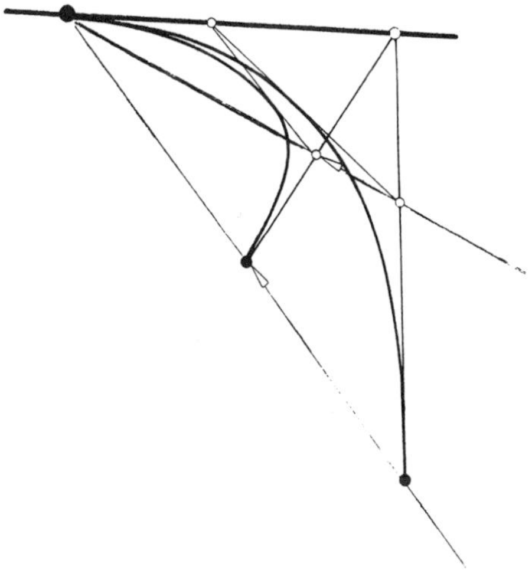

Figure 6.9. G^3 continuity: the center of the central collineation is at $\underline{\mathbf{b}}_0$.

Both kinds of geometric continuity can be used to refer to composite conics: just take a G^2 or G^3 piecewise conic and change one of its segments according to the above constructions. The resulting piecewise curve will still be G^2 or G^3.

6.8 Problems

1. Verify (6.10) for the case of a quarter circle in rational Bézier form. Then prove the general statement.

2. Devise a different system for estimating the $\mathbf{b}_{2i}$ from given $\mathbf{b}_{2i\pm1}$ in Section 6.5, using the idea of G^2 conic splines.

3. Numerically, (6.13) will rarely yield $I = 1$. Discuss what tolerances might be employed in determining if *one* conic can solve the G^2 interpolation problem.

7

Rational Bézier Curves

Bézier curves may be viewed as the backbone of all piecewise polynomial curve schemes; their rational counterparts enjoy a similar role for piecewise rational curves. Every piecewise polynomial or rational curve may be broken down into a collection of Bézier curves, and thus their study is crucial to the understanding of the whole field of NURBS.

Historically, Bézier curves – in their polynomial, or integral, version – are due to P. de Casteljau. In 1957, he invented what is now known as the de Casteljau algorithm, laid down in an internal report of the Citroën automotive company. His algorithm relies on the concept of constant ratios, and thus is tied in closely with affine geometry. Rational Bézier curves, however, may be evaluated using the concept of cross ratios, the fundamental invariant of projective geometry.[1]

[1]But when I sent de Casteljau a preprint of [43], he was not too enthusiastic about the new version of his algorithm: he just did not like the idea of using cross ratios...

7.1 The Bernstein Form

Any n–th degree polynomial curve $\underline{\mathbf{b}}(t)$ in $\mathbb{P}^3$ may be expressed in Bernstein form:

$$\underline{\mathbf{b}}(t) = \sum_{i=0}^{n} \underline{\mathbf{b}}_i B_i^n(t); \quad \underline{\mathbf{b}}_i \in \mathbb{P}^3, \tag{7.1}$$

where the B_i^n are the *Bernstein polynomials*

$$B_i^n(t) = \binom{n}{i} t^i (1-t)^{n-i}; \quad i = 0, \ldots, n. \tag{7.2}$$

We note two important properties of Bernstein polynomials: they form a *partition of unity*:

$$\sum_{i=0}^{n} B_i^n(t) \equiv 1 \tag{7.3}$$

and they satisfy the recursion

$$B_i^n(t) = (1-t)B_{i-1}^{n-1}(t) + tB_i^{n-1}(t); \quad B_{-1}^n(t) = B_{n+1}^n(t) \equiv 0. \tag{7.4}$$

For more properties, see [47].

We may project (7.1) into affine three-space: we will then obtain a *rational Bézier curve* $\mathbf{b}(t)$. This follows the same development as that of rational quadratics in Section 5.1, and yields

$$\mathbf{b}(t) = \frac{\sum_{i=0}^{n} w_i \mathbf{b}_i B_i^n(t)}{\sum_{i=0}^{n} w_i B_i^n(t)}; \quad \mathbf{b}_i \in \mathbb{E}^3. \tag{7.5}$$

The w_i are now called *weights*. If they are all positive, the curve lies in the convex hull of the control polygon $\mathbf{b}_0, \ldots, \mathbf{b}_n$.[2] This is called the *convex hull property* of Bézier curves.

If one weight, w_k, say, is increased, the curve is "pulled" towards $\mathbf{b}_k$.

We can make a more precise statement as follows: let $\mathbf{b}(t)$ be a point on some rational Bézier curve, and let $\bar{\mathbf{b}}(t)$ be a point on a curve with the same polygon and set of weights, except that w_k is replaced by cw_k. Then the three points $\mathbf{b}(t), \bar{\mathbf{b}}(t)$, and $\mathbf{b}_k$ are collinear. This is seen easily in a projective context: then we have to show that $\det[\underline{\mathbf{b}}(t), \underline{\bar{\mathbf{b}}}(t), \underline{\mathbf{b}}_k] = 0$. Writing this out in detail, we have

---

[2]The convex hull of the polygon is defined as the set of all points $\mathbf{p}$ that can be written as $\mathbf{p} = \alpha_0 \mathbf{b}_0 + \ldots + \alpha_n \mathbf{b}_n$ with $\alpha_0 + \ldots + \alpha_n = 1$ and $\alpha_i \geq 0$.

$$\begin{aligned}\det[\underline{\mathbf{b}}(t),\ \bar{\underline{\mathbf{b}}}(t),\ \underline{\mathbf{b}}_k] &= \det[\sum_{\substack{i=0\\ i\neq k}}^{n} \underline{\mathbf{b}}_i B_i^n(t) + \underline{\mathbf{b}}_k B_k^n(t),\\ &\qquad \sum_{\substack{i=0\\ i\neq k}}^{n} \underline{\mathbf{b}}_i B_i^n(t) + c\underline{\mathbf{b}}_k B_k^n(t),\ \underline{\mathbf{b}}_k]\\ &= \det[\underline{\mathbf{b}}_k(1-c)B_k^n(t),\ \bar{\underline{\mathbf{b}}}(t),\ \underline{\mathbf{b}}_k]\\ &= 0,\end{aligned}$$

since the first and the last columns of the last determinant are multiples of each other.[3]

Collinearity is preserved by going from projective to affine space, and so we have that $\bar{\mathbf{b}}(t)$ is a barycentric combination of $\mathbf{b}(t)$ and $\mathbf{b}_k$. As w_k increases, $\bar{\mathbf{b}}(t)$ is "pulled towards" $\mathbf{b}_k$, whereas for decreasing w_k, it is "pushed away" from $\mathbf{b}_k$.

We thus have two different ways to design rational Bézier curves: we may move a control point $\underline{\mathbf{b}}_k$, or we may change a weight w_k. A third way is given by the introduction of *weight points*

$$\mathbf{q}_i = \frac{(w_i\mathbf{b}_i + w_{i+1}\mathbf{b}_{i+1})}{w_i + w_{i+1}}. \tag{7.6}$$

While the weights determine the weight points, the converse is also true: given the weight points, we may compute a set of corresponding weights by observing that

$$\mathrm{ratio}(\mathbf{b}_i, \mathbf{q}_i, \mathbf{b}_{i+1}) = \frac{w_{i+1}}{w_i};$$

refer to Equation (2.20) for the definition of the ratio of three collinear points. Thus changing a weight point (and recomputing the weights) may be also used as a design tool.

Figure 7.1 gives an impression of how rational Bézier curves may be used in a design situation. With nonrational curves, the standard interactive tool is the displacement of control points. This is still possible for rational curves, of course. Additional flexibility is gained by the ability to change the weights as well. This is in fact a different design tool: when a control point is moved, all curve points move parallel to its displacement vector. When a weight is changed, all curve points move towards the corresponding control point or away from it. It may be more intuitive for a designer to work with the weight points (geometric) than directly dealing with the weights (algebraic).

The weight points $\mathbf{q}_i$ should be located between $\mathbf{b}_i$ and $\mathbf{b}_{i+1}$: if they are not, some weights will become negative.

---

[3] A similar technique was used in Section 4.10.

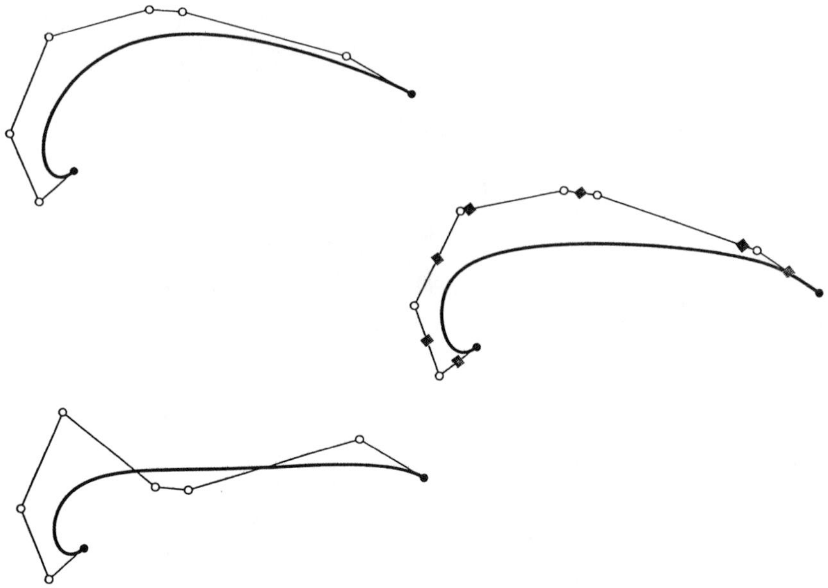

Figure 7.1. Designing with rational Bézier curves. Top: a nonrational curve of degree seven; middle: made rational. All weights equal one except $w_4 = w_5 = 0.1$. The weight points are shown as diamonds; bottom: two control points are moved (integral case).

The weight points may be used in the definition of a tighter convex hull for rational Bézier curves: in case of all weights w_i being positive, the curve lies in the convex hull of the point set $\{\mathbf{b}_0, \mathbf{q}_0, \mathbf{q}_1, \ldots, \mathbf{q}_{n-1}, \mathbf{b}_n\}$. This follows from the fact that the $\mathbf{q}_i$ are intermediate points in the subdivision process for $t = \frac{1}{2}$, see [48] and Section 7.9. The convex hull of the set $\{\mathbf{b}_0, \mathbf{q}_0, \mathbf{q}_1, \ldots, \mathbf{q}_{n-1}, \mathbf{b}_n\}$ is smaller than that of the control points. See Figure 7.2 for an illustration. Smaller convex hulls may be advantageous for algorithms like clipping or curve/plane intersections.

If some of the w_i differ in sign, the denominator

$$w(t) = \sum_{i=0}^{n} w_i B_i^n(t)$$

of (7.5) may become zero, causing a singularity. Geometrically, this means that the projective curve $\underline{\mathbf{b}}(t)$ crosses (or touches) the plane that is mapped to the affine plane at infinity, much in the same way in when we discussed the asymptotes of hyperbolas; see Section 3.5. The asymptote of a hyperbola is a straight line. Higher degree rational curves may have asymptotes that are themselves curves — this will happen if the weight function $w(t)$ has high order zeroes.

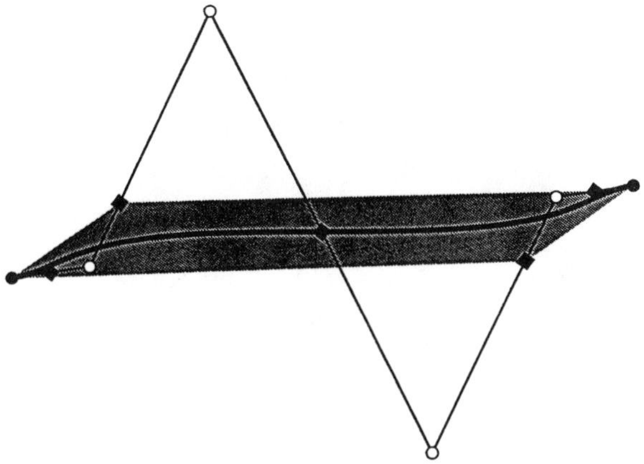

Figure 7.2. Weight points and convex hulls: if we use the weight points (diamonds), we may obtain a smaller size convex hull than by just using the control points.

7.2 The de Casteljau Algorithm

In projective space, $\underline{\mathbf{b}}(t)$ may be evaluated by a *projective de Casteljau algorithm:*

$$\underline{\mathbf{b}}_i^r(t) = (1-t)\underline{\mathbf{b}}_i^{r-1}(t) + t\underline{\mathbf{b}}_{i+1}^{r-1}(t); \begin{cases} r = 1, \ldots, n \\ i = 0, \ldots, n-r \end{cases} \quad (7.7)$$

with $\underline{\mathbf{b}}_i^0(t) = \underline{\mathbf{b}}_i$. Then $\underline{\mathbf{b}}_0^n(t) = \underline{\mathbf{b}}(t)$ is the point with parameter value t. Figure 7.3 illustrates the cubic case $n = 3$.[4] This de Casteljau algorithm is also defined in $\mathbb{E}^3$; there it yields polynomial Bézier curves — for more details, see [47].

If we introduce

$$\underline{\mathbf{q}}_i = (\underline{\mathbf{b}}_i + \underline{\mathbf{b}}_{i+1})/2,$$

we observe that

$$\operatorname{cr}(\underline{\mathbf{b}}_i, \underline{\mathbf{b}}_i^1, \underline{\mathbf{q}}_i, \underline{\mathbf{b}}_{i+1}) = \frac{t}{1-t}, \quad \text{all } i.$$

Just as we did to prove (4.6), we can define

$$\underline{\mathbf{q}}_i^r = (\underline{\mathbf{b}}_i^r \wedge \underline{\mathbf{b}}_{i+1}^r) \wedge (\underline{\mathbf{q}}_i^{r-1} \wedge \underline{\mathbf{q}}_{i+1}^{r-1}) \quad (7.8)$$

---

[4] The careful observer may notice that the collinear $\underline{\mathbf{b}}_i^r, \underline{\mathbf{b}}_i^{r+1}, \underline{\mathbf{b}}_{i+1}^r$ are not in the same ratio for all r: this is intentional, as ratios are not a projective concept.

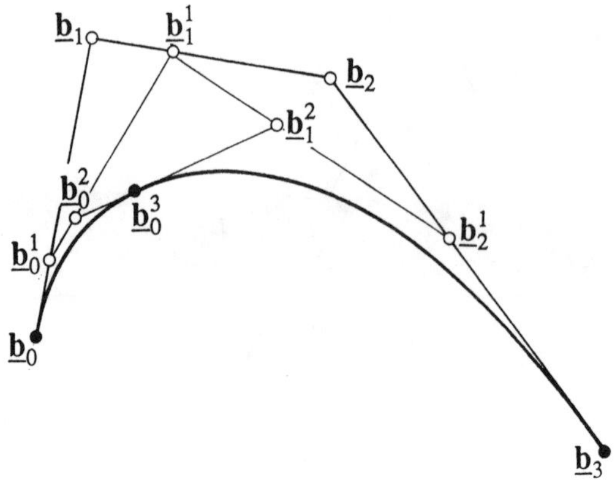

Figure 7.3. The projective de Casteljau algorithm: a point on a curve is constructed by repeated linear interpolation.

and obtain that

$$\underline{\mathbf{q}}_i^r = \frac{1}{2}\underline{\mathbf{b}}_i^r + \frac{1}{2}\underline{\mathbf{b}}_{i+1}^r. \tag{7.9}$$

As a consequence,

$$\operatorname{cr}(\underline{\mathbf{b}}_i^r, \underline{\mathbf{b}}_i^{r+1}, \underline{\mathbf{q}}_i^r, \underline{\mathbf{b}}_{i+1}^r) = \frac{t}{1-t}. \tag{7.10}$$

This shows that the projective de Casteljau algorithm (7.7) is projectively invariant: the point $\underline{\mathbf{b}}_0^n(t)$ can be found by intersections and cross ratio constructions only. The points $\underline{\mathbf{q}}_i$ are, of course, the projective preimages of the weight points $\mathbf{q}_i$.

We could also project every step of the projective de Casteljau algorithm into $\mathbb{E}^3$ and obtain the *rational de Casteljau algorithm*:

$$\mathbf{b}_i^r(t) = \frac{(1-t)w_i^{r-1}\mathbf{b}_i^{r-1}(t) + tw_{i+1}^{r-1}\mathbf{b}_{i+1}^{r-1}(t)}{w_i^r(t)} \qquad \begin{cases} r = 1, \ldots, n \\ i = 0, \ldots, n-r \end{cases} \tag{7.11}$$

with $\mathbf{b}_i^0(t) = \mathbf{b}_i$ and $w_i^r(t) = (1-t)w_i^{r-1}(t) + tw_{i+1}^{r-1}(t)$. Then $\mathbf{b}_0^n(t) = \mathbf{b}(t)$ is the point with parameter value t. Here, care must be taken in the case of t being outside the interval $[0,1]$, as some of the intermediate values may be vectors![5] Example 7.1 illustrates this: the projective point $\underline{\mathbf{b}}_2^1$ projects to an affine vector, and thus would need special case handling. It is best, therefore, to carry out all computations in projective space.

[5]The intermediate points could also be *close* to being vectors, i.e., having a very small third component. This may cause numerical problems.

The following is a de Casteljau scheme for the control polygon

$$\begin{bmatrix} -1 \\ 1 \end{bmatrix}, \begin{bmatrix} -1 \\ 0 \end{bmatrix}, \begin{bmatrix} 1 \\ 0 \end{bmatrix}, \begin{bmatrix} 1 \\ 1 \end{bmatrix},$$

weights $1, 2, 2, 1$ and $t = 2$. The computation is carried out in the projective plane.

$$\begin{array}{cccc}
\begin{bmatrix} -1 \\ 1 \\ 1 \end{bmatrix} & & & \\
\begin{bmatrix} -2 \\ 0 \\ 2 \end{bmatrix} & \begin{bmatrix} -3 \\ -1 \\ 3 \end{bmatrix} & & \\
\begin{bmatrix} 2 \\ 0 \\ 2 \end{bmatrix} & \begin{bmatrix} 6 \\ 0 \\ 2 \end{bmatrix} & \begin{bmatrix} 15 \\ 1 \\ 1 \end{bmatrix} & \\
\begin{bmatrix} 1 \\ 1 \\ 1 \end{bmatrix} & \begin{bmatrix} 0 \\ 2 \\ 0 \end{bmatrix} & \begin{bmatrix} -6 \\ 4 \\ -2 \end{bmatrix} & \begin{bmatrix} -27 \\ 7 \\ -5 \end{bmatrix}.
\end{array}$$

The last entry projects to the affine point $\mathbf{b}(2) = [27/5, -7/5]^{\mathrm{T}}$. Note that $\underline{\mathbf{b}}_2^1$'s third coordinate does not pose a problem at all!

Example 7.1. An example of the de Casteljau algorithm.

Finally, we present a different version of the rational de Casteljau algorithm described by Boehm [23]. It is sufficient to explain the basic idea for the case $n = 2$. There, we have (cf. Figure 7.4):

$$\begin{aligned}
\mathbf{b}_0^2 &= \frac{(1-t)w_0^1}{w_0^2}\mathbf{b}_0^1 + \frac{tw_1^1}{w_0^2}\mathbf{b}_1^1 =: (1-\beta)\mathbf{b}_0^1 + \beta\mathbf{b}_1^1, \\
\mathbf{b}_0^1 &= \frac{(1-t)w_0}{w_0^1}\mathbf{b}_0 + \frac{tw_1}{w_0^1}\mathbf{b}_1 =: (1-\alpha_0)\mathbf{b}_0^1 + \alpha_0\mathbf{b}_1^1, \\
\mathbf{b}_1^1 &= \frac{(1-t)w_1}{w_1^1}\mathbf{b}_1 + \frac{tw_2}{w_1^1}\mathbf{b}_2 =: (1-\alpha_1)\mathbf{b}_1 + \alpha_1\mathbf{b}_2.
\end{aligned}$$

It is straightforward to verify that

$$\frac{\beta}{1-\beta} = \frac{\alpha_0}{1-\alpha_1}. \tag{7.12}$$

This observation may be used to reformulate the rational de Casteljau algorithm: once we have computed the $\alpha_i; i = 0, \ldots, n-1$, we may continue

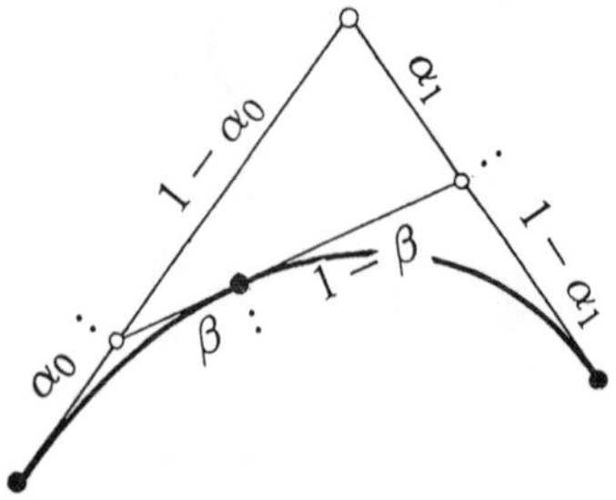

Figure 7.4. The rational de Casteljau algorithm: some of the involved barycentric combinations.

with barycentric combinations that use only these terms.[6] The following scheme shows, for $n = 4$, the involved ratios.

$$\begin{array}{llll} \alpha_0/(1-\alpha_0) & & & \\ \alpha_1/(1-\alpha_1) & \alpha_0/(1-\alpha_1) & & \\ \alpha_2/(1-\alpha_2) & \alpha_1/(1-\alpha_2) & \alpha_0/(1-\alpha_2) & \\ \alpha_3/(1-\alpha_3) & \alpha_2/(1-\alpha_3) & \alpha_1/(1-\alpha_3) & \alpha_0/(1-\alpha_3). \end{array} \tag{7.13}$$

In case of all $w_i = 1$, all these ratios become the familiar $t/(1-t)$ from the integral de Casteljau algorithm.

7.3 Degree Elevation

A projective Bézier curve $\underline{\mathbf{b}}^n(t) = \sum_{i=0}^n \underline{\mathbf{b}}_i B_i^n(t)$ of degree n may be expressed as a Bézier curve $\sum_{i=0}^{n+1} \underline{\mathbf{b}}_i^* B_i^{n+1}$ of degree $n+1$. The control points $\underline{\mathbf{b}}_i^*$ are given by

$$\underline{\mathbf{b}}_i^* = \frac{i}{n+1}\underline{\mathbf{b}}_{i-1} + (1 - \frac{i}{n+1})\underline{\mathbf{b}}_i; \quad i = 0, \ldots, n+1. \tag{7.14}$$

This is seen by multiplying $\sum_{i=0}^n \underline{\mathbf{b}}_i B_i^n(t)$ by $[t + (1-t)]$ and comparing powers of t^i and $(1-t)^{n+1-i}$.

We can project this into affine space, and obtain the *degree elevation*

[6] We should note that this reformulation is more of a geometric than of a computational nature: the algorithm proceeds as before!

formulas for rational Bézier curves:

$$\mathbf{b}_i^* = \frac{i w_{i-1}\mathbf{b}_{i-1} + (n+1-i)w_i\mathbf{b}_i}{i w_{i-1} + (n+1-i)w_i}; \quad i = 0,\ldots,n+1, \tag{7.15}$$

$$w_i = i w_{i-1} + (n+1-i)w_i. \tag{7.16}$$

Figure 7.5 gives an example of a conic that is degree elevated to rational cubic form.

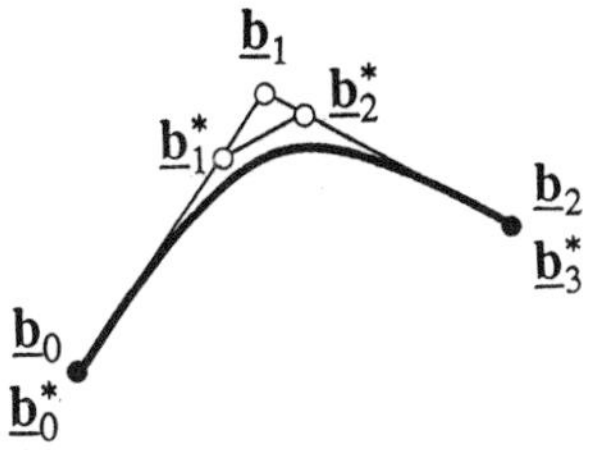

Figure 7.5. Rational degree elevation: a rational quadratic is degree elevated to rational cubic.

We may degree elevate a rational Bézier curve repeatedly, resulting in a sequence of weights and control polygons. Provided that all weights are of the same sign, the control polygons will converge to the curve. This follows from the corresponding results for polynomial curves; see [42] or [47].

We proved (7.14) by multiplying the equation of a Bézier curve by the "harmless" constant 1, written as $1 = [t + (1-t)]$. But we may multiply $\underline{\mathbf{b}}(t)$ by any other nonzero function $\rho(t)$ and would not change its shape! The function ρ may be polynomial, thus raising the degree of the original curve. If ρ has zeroes, we will introduce base points; see Section 7.7.

7.4 Degree Reduction

In general, we will not be able to lower the degree of a given rational Bézier curve. For example, a cubic, which might have an inflection point, cannot be written as a rational quadratic, because a conic cannot have an inflection point. The process of degree reduction will therefore be approximative. For the special case of approximating a rational cubic by a rational quadratic, see Section 8.2.

We will choose to perform degree reduction in projective space – there, we only have to deal with polynomials. Our aim is therefore the following: given a Bézier curve $\underline{\mathbf{b}}(t)$ with control vertices $\underline{\mathbf{b}}_i; i = 0,\ldots,n$, find a Bézier curve $\hat{\underline{\mathbf{b}}}(t)$ with control vertices $\hat{\underline{\mathbf{b}}}_i; i = 0, \ldots, n-1$ that approximates the first curve. Several methods exist for this purpose, although they have all been developed in a Euclidean context. They can all be cast in the following setting.

Let us pretend that the $\underline{\mathbf{b}}_i$ were obtained from the $\hat{\underline{\mathbf{b}}}_i$ by the process of degree elevation (this is not true, in general, but makes a good working assumption). Then they would be related by

$$\underline{\mathbf{b}}_i = \frac{i}{n}\hat{\underline{\mathbf{b}}}_{i-1} + \frac{n-i}{n}\hat{\underline{\mathbf{b}}}_i; \quad i = 0, 1, \ldots, n. \tag{7.17}$$

This equation can be used to derive two recursive formulas for the generation of the $\hat{\underline{\mathbf{b}}}_i$ from the $\underline{\mathbf{b}}_i$:

$$\hat{\underline{\mathbf{b}}}_i^{\text{left}} = \frac{n\underline{\mathbf{b}}_i - i\hat{\underline{\mathbf{b}}}_{i-1}^{\text{left}}}{n-i}; \quad i = 0, 1, \ldots, n-1 \tag{7.18}$$

and

$$\hat{\underline{\mathbf{b}}}_{i-1}^{\text{right}} = \frac{n\underline{\mathbf{b}}_i - (n-i)\hat{\underline{\mathbf{b}}}_i^{\text{right}}}{i}; \quad i = n, n-1, \ldots, 1. \tag{7.19}$$

The $\hat{\underline{\mathbf{b}}}_i^{\text{left}}$ perform satisfactorily near $\underline{\mathbf{b}}_0$, whereas the $\hat{\underline{\mathbf{b}}}_{i-1}^{\text{right}}$ do well near $\underline{\mathbf{b}}_n$. It seems a good idea, therefore, to *blend* the two sets in order to obtain a final control net $\hat{\underline{\mathbf{b}}}_i$:

$$\hat{\underline{\mathbf{b}}}_i = (1-\alpha_i)\hat{\underline{\mathbf{b}}}_i^{\text{left}} + \alpha_i\hat{\underline{\mathbf{b}}}_i^{\text{right}}. \tag{7.20}$$

R. Forrest [58] has suggested to set $\alpha_i = 0$ for $i < n/2$ and $\alpha_i = 1$ for $i > n/2$, with a case distinction for $\alpha_{n/2}$ when n is odd. Another choice with reasonable outcome is $\alpha_i = i/(n-1)$. An "optimal" choice is given by

$$\alpha_i = 2^{1-2n}\sum_{j=0}^{i}\binom{2n}{2j}; \quad i = 0, \ldots, n-1. \tag{7.21}$$

In Euclidean space, this choice ensures that we obtain the best approximation (in the Chebychev sense) of degree $n-1$ to the given polynomial of degree n; see Eck[41]. In projective space, such claims are meaningless, but we should still expect (7.21) to perform well.

The first appearance of degree reduction is in Forrest [58]. Eck's geometric treatment is similar to more algebraic methods by Watkins and Worsey [117] and Lachance [75].

7.5 Reparametrization

We have seen that an arc of a conic does not have a unique parametric representation: depending on the choice of the shoulder tangent, each weight w_i could be multiplied by the power c^i of a nonzero constant c: this would not change the geometry of the curve, but it would change how the parameter traverses it. See Section 4.9 for details. The same principle also serves to reparametrize Bézier curves, in projective space as well after being projected into affine space.

We can therefore write a projective Bézier curve in the following two ways:

$$\sum_{i=0}^{n} \underline{\mathbf{b}}_i B_i^n(t) = \sum_{i=0}^{n} c^i \underline{\mathbf{b}}_i B_i^n(\hat{t}),$$

which means that a rational Bézier curve can be written as

$$\frac{\sum_{i=0}^{n} w_i \mathbf{b}_i B_i^n(t)}{\sum_{i=0}^{n} w_i B_i^n(t)} = \frac{\sum_{i=0}^{n} c^i w_i \mathbf{b}_i B_i^n(\hat{t})}{\sum_{i=0}^{n} c^i w_i B_i^n(\hat{t})}.$$

Reparametrization of a rational Bézier curve changes the rate in which the parameter t traverses the curve; Figure 7.6 illustrates.

After a curve has been reparametrized, it will have new weight points $\hat{\mathbf{q}}_i$. Just as in the conic case, the old and new weight points are related by

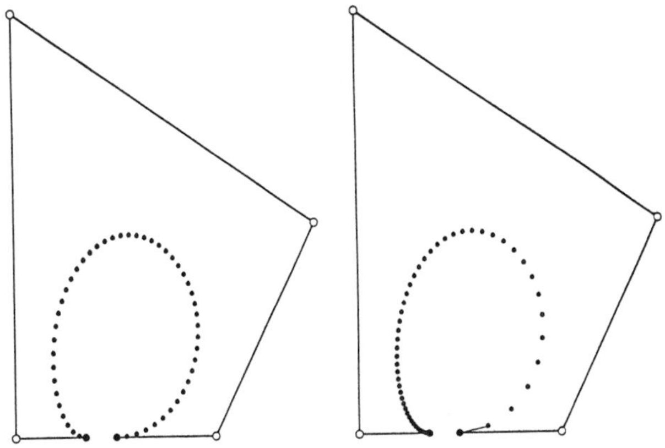

Figure 7.6. Reparametrization: a rational Bézier curve is sampled at 50 equal parameter intervals. Left, $c = 1$; right: c=1/3.

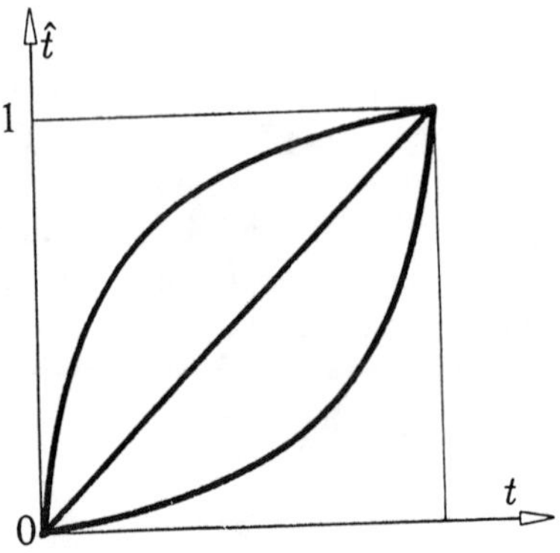

Figure 7.7. Reparametrization: the relationship between old and new parameter depending on the constant c.

a constant cross ratio relation:

$$\mathrm{cr}(\mathbf{b}_i, \mathbf{q}_i, \hat{\mathbf{q}}_i, \mathbf{b}_{i+1}) = \frac{1}{c}. \tag{7.22}$$

The old and new parameters of the curve are related by the same rational linear, or Moebius, transformation (4.36) that was used for the conic case:

$$\hat{t} = \frac{ct}{1 + (c-1)t}. \tag{7.23}$$

A graph of the rational linear function (7.23) is shown in Figure 7.7. [7]

A rational Bézier curve with positive weights may be reparametrized so that the end weights are unity: this is achieved by selecting

$$c = \sqrt[n]{\frac{w_n}{w_0}}.$$

Curves in this form are said to be in *standard form*

[7] It is interesting to note that another, very similar class, of reparametrizations is used in grid generation for finite element analysis: there, one encounters transformations of the form

$$\hat{t} = \frac{e^{kt} - 1}{e^k - 1}. \tag{7.24}$$

The graphs of these functions, for varying k, looks much the same as those in 7.7. For our purposes, it is important to note that (7.23) produces yet another rational curve, whereas (7.24) takes us out of that class.

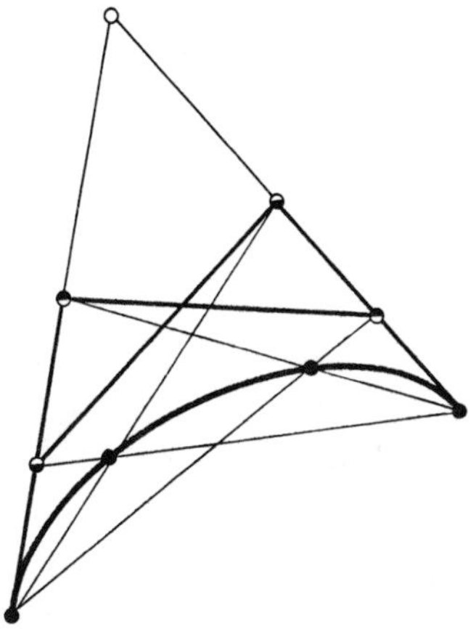

Figure 7.8. Rational ambiguities: a rational quadratic may be written as two different rational cubics using degree elevation and reparametrization.

Reparametrization and degree elevation together can cause an effect that cannot be found in polynomial curves. Take a rational cubic, for example, and degree elevate it. We could take that same rational cubic, reparametrize it, and then degree elevate it. In this way, we obtain two different rational quartic representations of the same curve! See Figure 7.8 for an illustration. The polynomial case is quite different: two polynomial Bézier curves of the same degree are equal if and only if their control polygons are the same.

7.6 Nonparametric Curves

A functional, or nonparametric, integral Bézier curve is defined by

$$y = \sum_{i=0}^{n} b_i B_i^n(x).$$

It is represented easily in terms of a parametric Bézier curve: its Bézier points are given by $\mathbf{b}_i = [i/n, b_i]^{\mathrm{T}}$. This works since the $x-$coordinates satisfy

$$x = \sum_{i=0}^{n} \frac{i}{n} B_i^n(x). \tag{7.25}$$

The situation in the rational case is somewhat more complicated. Let

$$y = \frac{\sum w_i b_i B_i^n(x)}{\sum w_i B_i^n(x)} \tag{7.26}$$

be a rational function of x. Can we write it as a parametric rational Bézier curve, and if so, what are its Bézier points $\mathbf{b}_i$ and the corresponding weights v_i? Written in parametric form, (7.26) takes the form

$$\begin{bmatrix} x(u) \\ y(u) \end{bmatrix} = \begin{bmatrix} u \\ \dfrac{\sum w_i b_i B_i^n(u)}{\sum w_i B_i^n(u)} \end{bmatrix} = \begin{bmatrix} \dfrac{u \sum w_i B_i^n(u)}{\sum w_i B_i^n(u)} \\ \dfrac{\sum w_i b_i B_i^n(u)}{\sum w_i B_i^n(u)} \end{bmatrix}.$$

Using the identity $uB_i^n(u) = \frac{i+1}{n+1} B_{i+1}^{n+1}(u)$ and degree elevation, we obtain

$$\begin{bmatrix} x(u) \\ y(u) \end{bmatrix} = \frac{\sum_{i=0}^{n+1} v_i \mathbf{b}_i B_i^{n+1}(u)}{\sum_{i=0}^{n+1} v_i B_i^{n+1}(u)},$$

where

$$\mathbf{b}_i = \frac{1}{v_i} \begin{bmatrix} iw_{i-1} \\ iw_{i-1}b_{i-1} + (n+1-i)w_i b_i \end{bmatrix}; \quad i = 0, \ldots, n+1$$

and

$$v_i = iw_{i-1} + (n+1-i)w_i; \quad i = 0, \ldots, n+1.$$

In the integral case, the abscissae i/n of the Bézier points of a functional Bézier curve do not depend on the function; now they depend on the weights of the function under consideration. Thus a rational function $y = a(x)/b(x)$ with both a and b polynomials of degree n has a rational parametric representation of degree $n+1$.

The abscissae iw_{i-1}/v_i are in the range $[0, 1]$. But note that they do not necessarily have to be increasing! For example, if the original weight sequence $\{w_i\}$ is $1, 10, 1, 10$, then the abscissae sequence $\{iw_{i-1}/v_i\}$ is $0, 1/31, 10/11, 3/13, 1$. See Example 7.2 and the Problem section for more cases.

7.7 Base Points

What happens if all components of a projective Bézier curve vanish simultaneously, yielding an expression of the form $\mathbf{b}(t) = \underline{\mathbf{0}}$, which is clearly meaningless? In this situation, the corresponding parameter value t is

Using degree elevation, the function $y = 1/(1+x)$ may be written as

$$y = \frac{1 \cdot 1 \cdot B_0^1(x) + 2 \cdot \frac{1}{2} \cdot B_1^1(x)}{1 \cdot B_0^1(x) + 2 \cdot B_1^1(x)}.$$

We may write the segment over $0 \le x \le 1$ as a rational quadratic; its Bézier points are

$$\mathbf{b}_i = \begin{bmatrix} 0 \\ 1 \end{bmatrix}, \begin{bmatrix} 1/3 \\ 2/3 \end{bmatrix}, \begin{bmatrix} 1 \\ 1/2 \end{bmatrix},$$

and the corresponding weights are $v_i = 2, 3, 4$. We may standardize this curve and obtain weights $\hat{v}_i = 1, \frac{3}{4}\sqrt{2}, 1$. The middle weight is larger than one, indicating that our example function represents a hyperbola.

Example 7.2. A parametric representation of a rational linear function.

called a *base point.* Thus a base point is a parameter value that does not map to a unique point in the range.

Let us study the case $t = 0$ instead of the general case. This is no restriction since we may employ subdivision (see Section 7.9) to make any point on a curve correspond to $t = 0$. If we have a base point for $t = 0$, then it follows that $\underline{\mathbf{b}} = \underline{\mathbf{0}}$ and our Bézier curve reduces to

$$\begin{aligned}
\underline{\mathbf{b}}(t) &= \sum_{i=1}^{n} \underline{\mathbf{b}}_i \binom{n}{i} t^i (1-t)^{n-i} \\
&= t \sum_{i=1}^{n} \underline{\mathbf{b}}_i \binom{n}{i} t^{i-1} (1-t)^{n-i} \\
&\hat{=} \sum_{i=0}^{n-1} \underline{\mathbf{b}}_{i+1} \binom{n-1}{i+1} B_i^{n-1}(t).
\end{aligned}$$

This has to be interpreted as follows: the base point reduces the degree of the curve by one. For $t = 0$, the curve now starts at $\underline{\mathbf{b}}_1$ instead of $\underline{\mathbf{b}}_0$.

For any curve in projective space, base points may be introduced by multiplying by an arbitrary function $\phi(t)$ – recall that this does not change the shape of the curve. So replacing $\underline{\mathbf{b}}(t)$ by $\phi(t)\underline{\mathbf{b}}(t)$ results in base points at every zero of $\phi(t)$. If ϕ is a rational polynomial, then so is $\phi(t)\underline{\mathbf{b}}(t)$, but now with base points present. Such polynomials are called *improperly parametrized*; Sederberg [104] discusses how to detect and to remove such parametrizations.

Another way of introducing a base point is by *degree elevation* . If we write a polynomial curve in projective space in monomial form: $\underline{\mathbf{x}}(t) = \underline{\mathbf{a}}_0 + \underline{\mathbf{a}}t + \ldots + \underline{\mathbf{a}}_n t^n$, then degree elevation amounts to formally adding a term in t^{n+1}, i.e., $\underline{\mathbf{x}}(t) = \underline{\mathbf{a}}_0 + \underline{\mathbf{a}}t + \ldots + \underline{\mathbf{a}}_n t^n + \underline{\mathbf{0}}t^{n+1}$. As we let t tend to infinity, the curve will tend to $\underline{\mathbf{0}}$, the coefficient of the highest power of t.[8] In this sense, degree elevation amounts to the introduction of a base point at $t = \infty$.

7.8 Derivatives

The derivative of a rational function seemingly involves the quotient rule; but we will avoid its use just as we did in the conic case. So we define $\mathbf{p}$ by

$$\mathbf{p}(t) = w(t)\mathbf{b}(t), \tag{7.27}$$

where $w(t)$ is the denominator of the rational Bézier curve $\mathbf{b}(t)$. Following the steps in Section 5.4, we obtain for the derivative $\dot{\mathbf{b}}(t)$:

$$\dot{\mathbf{b}}(t) = \frac{1}{w(t)}[\dot{\mathbf{p}}(t) - \dot{w}(t)\mathbf{b}(t)], \tag{7.28}$$

the dot denoting differentiation with respect to t. For the first derivatives at the endpoints of a rational Bézier curve, we find

$$\dot{\mathbf{b}}(0) = \frac{nw_1}{w_0}\Delta\mathbf{b}_0 \tag{7.29}$$

and

$$\dot{\mathbf{b}}(1) = \frac{nw_{n-1}}{w_n}\Delta\mathbf{b}_{n-1}. \tag{7.30}$$

For higher derivatives, we differentiate (7.27) r times:

$$\mathbf{p}^{(r)}(t) = \sum_{j=0}^{r}\binom{r}{j}w^{(j)}(t)\mathbf{b}^{(r-j)}(t).$$

We can solve for $\mathbf{b}^{(r)}(t)$:

$$\mathbf{b}^{(r)}(t) = \frac{1}{w(t)}[\mathbf{p}^{(r)} - \sum_{j=1}^{r}\binom{r}{j}w^{(j)}(t)\mathbf{b}^{(r-j)}(t)]. \tag{7.31}$$

This is a recursive formula for the rth derivative of a rational Bézier curve. It only involves taking derivatives of polynomial curves, thus avoiding the quotient rule all together.

---

[8] This also follows from the fact that the nth derivative of an nth degree polynomial agrees with its value for $t = \infty$.

It is also possible to compute derivatives using the rational de Casteljau algorithm(7.11). We then obtain

$$\dot{\mathbf{b}}(t) = n\frac{w_0^{n-1}w_1^{n-1}}{[w_0^n]^2}\left[\mathbf{b}_1^{n-1} - \mathbf{b}_0^{n-1}\right]. \tag{7.32}$$

For a proof, see Floater [54]. Note that we do have to carry out the full rational de Casteljau algorithm; we could compute everything in projective space, up to $r = n - 1$, that is. Then we project into affine space so that we can apply (7.32).

For some algorithms, it is useful to know an upper bound for the derivative (the hodograph) of a curve. Sederberg and Wang [108] give the bound

$$||\dot{\mathbf{b}}(t)|| \le n\frac{\max\{w_i\}}{\min\{w_i\}}\max_{0\le i,j\le n}\{||\mathbf{b}_i - \mathbf{b}_j||\} \tag{7.33}$$

for $0 \le t \le 1$. Floater [54] gives

$$||\dot{\mathbf{b}}(t)|| \le n\left[\frac{\max\{w_i\}}{\min\{w_i\}}\right]^2\max_{0\le i<n}\{||\mathbf{b}_{i+1} - \mathbf{b}_i||\} \tag{7.34}$$

and notes that neither bound is sharper than the other. Of course, (7.34) is much faster to compute!

At the endpoints of a curve, derivatives are used to describe the curve's geometry. In terms of projective geometry, the *tangent* at $t = 0$ is given by the line through $\underline{\mathbf{b}}_0$ and $\underline{\mathbf{b}}_1$; the *osculating plane* at $t = 0$ is given by the plane $< \underline{\mathbf{b}}_0, \underline{\mathbf{b}}_1, \underline{\mathbf{b}}_2 >$, with analogous results for $t = 1$. For the special case of a cubic, we see that the Bézier point $\underline{\mathbf{b}}_1$ is the intersection of the osculating plane at $t = 1$ and the tangent at $t = 0$, and that $\underline{\mathbf{b}}_2$ is the intersection of the osculating plane at $t = 0$ and the tangent at $t = 1$.

The concepts of tangents and osculating planes are projectively invariant; thus the same statements hold for the rational case.

7.9 Blossoms

We already introduced the concepts of blossoms for projective quadratic curves in Section 4.8. Quite analogously, we now define the blossom $\underline{\mathbf{b}}[t_1, \ldots, t_n]$ by applying the de Casteljau algorithm to the control polygon $\underline{\mathbf{b}}_0, \ldots, \underline{\mathbf{b}}_n$, but with a different argument t_i for each level of the algorithm. From the discussion in Section 4.8, it is clear that the value of $\underline{\mathbf{b}}[t_1, \ldots, t_n]$ does not depend on the order in which the t_i are fed into the algorithm. The blossom $\underline{\mathbf{b}}[t_1, \ldots, t_n]$ is thus a symmetric function of n variables. If all variables are equal: $t_1 = \ldots = t_n = t$, then the blossom agrees with the curve. It is also clear from its definition that the blossom is multilinear:

$$\underline{\mathbf{b}}[\ldots,(1-\alpha)r+\alpha s,\ldots] = (1-\alpha)\underline{\mathbf{b}}[\ldots,r,\ldots]+\alpha\underline{\mathbf{b}}[\ldots,s,\ldots]. \quad (7.35)$$

What is not immediately clear, and in fact requires a thorough proof, is the fact that the blossom is uniquely associated with the given curve — it does not depend on its initial form, Bézier or other.

The original de Casteljau algorithm, with its intermediate points $\underline{\mathbf{b}}_i^r$, may be rewritten in blossom form; we give the cubic example:

$$\begin{array}{llll} \underline{\mathbf{b}}_0 & & & \\ \underline{\mathbf{b}}_1 & \underline{\mathbf{b}}_0^1 & & \\ \underline{\mathbf{b}}_2 & \underline{\mathbf{b}}_1^1 & \underline{\mathbf{b}}_0^2 & \\ \underline{\mathbf{b}}_3 & \underline{\mathbf{b}}_2^1 & \underline{\mathbf{b}}_1^2 & \underline{\mathbf{b}}_0^3 \end{array} \quad = \quad \begin{array}{llll} \underline{\mathbf{b}}[0,0,0] & & & \\ \underline{\mathbf{b}}[0,0,1] & \underline{\mathbf{b}}[0,0,t] & & \\ \underline{\mathbf{b}}[0,1,1] & \underline{\mathbf{b}}[0,1,t] & \underline{\mathbf{b}}[0,t,t] & \\ \underline{\mathbf{b}}[1,1,1] & \underline{\mathbf{b}}[1,1,t] & \underline{\mathbf{b}}[1,t,t] & \underline{\mathbf{b}}[t,t,t]. \end{array}$$

In general, the $\underline{\mathbf{b}}_i^r(t)$ are written in blossom form as

$$\underline{\mathbf{b}}_i^r(t) = \underline{\mathbf{b}}[0^{<n-r-i>}, t^{<r>}, 1^{<i>}].$$

The blossom can also be used to *subdivide* a Bézier curve: if we are interested in the control points $\underline{\mathbf{c}}_i$ of the curve segment corresponding to $[0,b]$, all we have to do is to evaluate the blossom:

$$\underline{\mathbf{c}}_i = \underline{\mathbf{b}}[0^{<n-i>}, b^{<i>}]. \quad (7.36)$$

A special case ($b = \frac{1}{2}$) was already presented for conics in Section 5.3.

If we are interested in the control points $\underline{\mathbf{c}}_i$ corresponding to an arbitrary interval $[a,b]$, we proceed in the same way:

$$\underline{\mathbf{c}}_i = \underline{\mathbf{b}}[a^{<n-i>}, b^{<i>}]. \quad (7.37)$$

Blossoms may also be used to compute derivatives of Bézier curves in projective space. The following is the natural generalization of the conic case (4.31):

$$\frac{\mathrm{d}^r}{\mathrm{d}t^r}\underline{\mathbf{b}}(t) \hat{=} \underline{\mathbf{b}}[\infty^{<r>}, t^{<n-r>}]. \quad (7.38)$$

Special blossoms, called *osculants* may be defined at the curve's end-points:

$$\underline{\mathbf{o}}_l^r(t) = \underline{\mathbf{b}}[0^{<n-r>}, t^{<r>}], \quad \underline{\mathbf{o}}_r^r(t) = \underline{\mathbf{b}}[1^{<n-r>}, t^{<r>}]. \quad (7.39)$$

These are curves of degree r that agree in all r derivatives with those of the curve at $t = 0$ or $t = 1$, respectively. For $r = 1$, they are simply tangent

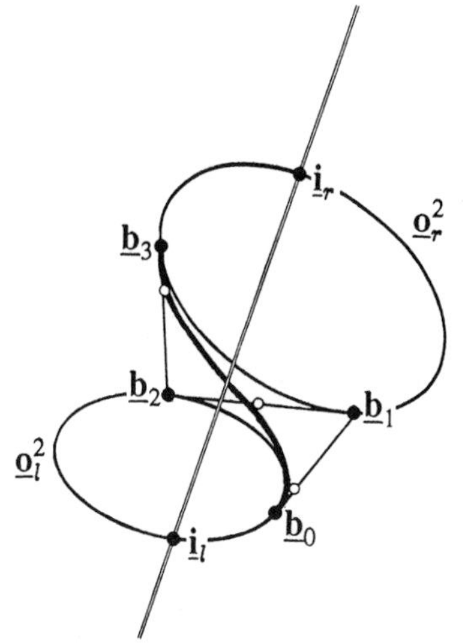

Figure 7.9. Osculants: the osculants at the endpoints of a cubic are shown.

lines. For $r = 2$, they are conics. The plane that is defined by such a conic osculant is called the *osculating plane* of the curve.

Figure 7.9 shows two osculants $\underline{\mathbf{o}}_l^2$ and $\underline{\mathbf{o}}_r^2$ to a cubic. Note that the line $\underline{\mathbf{o}}_l^2(\infty) \wedge \underline{\mathbf{o}}_r^2(\infty)$ contains all second derivatives to the cubic (where $\underline{\mathbf{o}}_1^2(\infty) = \underline{\mathbf{i}}_1$ and $\underline{\mathbf{o}}_r^2(\infty) = \underline{\mathbf{i}}_r$).

Osculants may be defined for arbitrary parameter values; at parameter value s, we then have the osculant $\underline{\mathbf{b}}[s^{<n-r>}, t^{<r>}]$. For the special case of $s = \infty$, the osculants agree with derivatives.

The idea of osculants goes back to Jolles [73]; see [25].

7.10 Control Vectors

In Section 5.5, we encountered control vectors (also known as infinite control points) as the limiting case of parallel tangents to a conic. The resulting rational curve representation contained both points and vectors. We can devise a similar form for rational Bézier curves (first suggested by K. Vesprille [115]). They will be of the form

$$\mathbf{b}(t) = \frac{\sum_{\text{points}} w_i \mathbf{b}_i B_i^n(t) + \sum_{\text{vectors}} \mathbf{v}_i B_i^n(t)}{\sum_{\text{points}} w_i B_i^n(t)}; \tag{7.40}$$

see also [52]. The control vectors $\mathbf{v}_i$ do not have weights in this form; we may multiply each of them by a factor, and the curve will change accordingly. Note that at least one of the point weights w_i must be nonzero for (7.40) to be meaningful – otherwise a point (namely $\mathbf{b}(t)$) would have to equal a vector!

For theoretical purposes, the control vector form is attractive and has received some attention in the literature; see [92], [90], [91]. But it is not as useful in practical situations: as in the conic case, we have lost the convex hull property, and evaluation of (7.40) will require special case treatment. Also, control vectors cannot be incorporated into the current IGES standard; see Section 15.1.

7.11 Duality

We derived Bézier curves as a generalization of the line conic construction from Section 4.2. Of course, we could also generalize the point conic construction from Section 4.3. In the planar case, this leads to what was called *dual Bézier curves* by J. Hoschek [68].[9] Further literature: Boddoluri and Ravani [19] and [50].

To define these curves, we are given a set of *control lines* $\underline{\mathbf{B}}_i$ in the projective plane. We define a curve by defining its tangent $\underline{\mathbf{B}}(t)$ for every parameter value t:

$$\underline{\mathbf{B}}(t) = \sum_{i=0}^{n} \underline{\mathbf{B}}_i(t). \tag{7.41}$$

We say that (7.41) defines a *control line* Bézier curve. (In this context, we shall call the "usual" Bézier curves *control point* Bézier curves.) Thus the curve is given as the envelope of its tangents. We see that the tangents at $t = 0$ and $t = 1$ are given by $\underline{\mathbf{B}}_0$ and $\underline{\mathbf{B}}_1$, respectively.

We can develop a de Casteljau algorithm for control line curves just as we could for control point curves. Any two subsequent lines $\underline{\mathbf{B}}_i$ and $\underline{\mathbf{B}}_{i+1}$ define a point $\underline{\mathbf{b}}_i = \underline{\mathbf{B}}_i \wedge \underline{\mathbf{B}}_{i+1}$, which in turn carries a pencil of lines through it. In each of these pencils, we can define a line $\underline{\mathbf{Q}}_i = \frac{1}{2}\underline{\mathbf{B}}_i + \frac{1}{2}\underline{\mathbf{B}}_{i+1}$; $i = 0, \ldots, n-1$. The de Casteljau algorithm now becomes:

$$\underline{\mathbf{B}}_i^r(t) = (1-t)\underline{\mathbf{B}}_i^{r-1}(t) + t\underline{\mathbf{B}}_{i+1}^{r-1}(t),$$

thus assuring that

$$\operatorname{cr}(\underline{\mathbf{B}}_i^{r-1}, \underline{\mathbf{B}}_i^r, \underline{\mathbf{Q}}_i^{r-1}, \underline{\mathbf{B}}_{i+1}^{r-1}) = \frac{t}{1-t},$$

[9] This wording is misleading: both concepts are dual to each other – neither is favored over the other.

where we have set, in analogy to (7.8):

$$\underline{\mathbf{Q}}_i^r = (\underline{\mathbf{B}}_i^r \wedge \underline{\mathbf{B}}_{i+1}^r) \wedge (\underline{\mathbf{Q}}_i^{r-1} \wedge \underline{\mathbf{Q}}_{i+1}^{r-1}).$$

This is, of course, exactly the dual of the de Casteljau algorithm (7.7).

For control point curves, the tangent corresponding to t is given by $\underline{\mathbf{T}}(t) = \underline{\mathbf{b}}(t) \wedge \dot{\underline{\mathbf{b}}}(t)$. For control line curves, we then have that the point corresponding to t is given as the intersection of two lines, namely $\underline{\mathbf{p}}(t) = \underline{\mathbf{B}}(t) \wedge \dot{\underline{\mathbf{B}}}(t)$. In particular, the endpoints of the curve are given by $\underline{\mathbf{B}}_0 \wedge \underline{\mathbf{B}}_1$ and $\underline{\mathbf{B}}_{n-1} \wedge \underline{\mathbf{B}}_n$, respectively. We may also express the point on the curve in terms of the de Casteljau algorithm: $\underline{\mathbf{p}}(t) = \underline{\mathbf{B}}_0^{n-1}(t) \wedge \underline{\mathbf{B}}_1^{n-1}(t)$.

If we are given a control point curve, defined by a set of control points $\underline{\mathbf{b}}_0, \ldots, \underline{\mathbf{b}}_n$, how can we express it as a control line curve, with control lines $\underline{\mathbf{B}}_0, \ldots, \underline{\mathbf{B}}_m$?[10] Instead of treating the general case, we first restrict ourselves to the cubic case $n = 3$. Recall that the tangent $\underline{\mathbf{T}}(t)$ at $\underline{\mathbf{b}}_0^3(t)$ is spanned by the intermediate control points $\underline{\mathbf{b}}_0^2(t)$ and $\underline{\mathbf{b}}_1^2(t)$:

$$\begin{aligned}
\underline{\mathbf{T}} &= \underline{\mathbf{b}}_0^2(t) \wedge \underline{\mathbf{b}}_1^2(t) \\
&= [(1-t)^2\underline{\mathbf{b}}_0 + 2t(1-t)\underline{\mathbf{b}}_1 + t^2\underline{\mathbf{b}}_2] \\
&\quad \wedge[(1-t)^2\underline{\mathbf{b}}_1 + 2t(1-t)\underline{\mathbf{b}}_2 + t^2\underline{\mathbf{b}}_3] \\
&= (1-t)^4\underline{\mathbf{b}}_0 \wedge \underline{\mathbf{b}}_1 \\
&\quad +2t(1-t)^3\underline{\mathbf{b}}_0 \wedge \underline{\mathbf{b}}_2 \\
&\quad +t^2(1-t)^2(\underline{\mathbf{b}}_0 \wedge \underline{\mathbf{b}}_3 + 3\underline{\mathbf{b}}_1 \wedge \underline{\mathbf{b}}_2) \\
&\quad +2t^3(1-t)\underline{\mathbf{b}}_1 \wedge \underline{\mathbf{b}}_3 \\
&\quad +t^4\underline{\mathbf{b}}_2 \wedge \underline{\mathbf{b}}_3.
\end{aligned}$$

Thus we now have a quartic control line curve whose control lines are given by

$$\begin{gathered}
\underline{\mathbf{b}}_0 \wedge \underline{\mathbf{b}}_1 \\
\tfrac{1}{2}\underline{\mathbf{b}}_0 \wedge \underline{\mathbf{b}}_2 \\
\tfrac{1}{6}(\underline{\mathbf{b}}_0 \wedge \underline{\mathbf{b}}_3 + 3\underline{\mathbf{b}}_1 \wedge \underline{\mathbf{b}}_2) \\
\tfrac{1}{2}\underline{\mathbf{b}}_1 \wedge \underline{\mathbf{b}}_3 \\
\underline{\mathbf{b}}_2 \wedge \underline{\mathbf{b}}_3.
\end{gathered}$$

Dually, we may convert a control line cubic to a control point quartic, using the same formalism as above.

In general, this conversion process will raise the degree from n to $2n-2$, considering that it involves terms of the form $\underline{\mathbf{T}}(t) = \underline{\mathbf{b}}_0^{n-1}(t) \wedge \underline{\mathbf{b}}_1^{n-1}(t)$ or $\underline{\mathbf{b}}(t) = \underline{\mathbf{B}}_0^{n-1}(t) \wedge \underline{\mathbf{B}}_1^{n-1}(t)$.

[10] Note that we do not know the degree m of the control line form yet!

Let us now turn to the 3-D case. The "usual" Bézier curves are again called control point curves; their duals are *control plane* curves of the form

$$\underline{\mathbf{B}}(t) = \sum_{i=0}^{n} \underline{\mathbf{B}}_i(t),, \tag{7.42}$$

but now with $\underline{\mathbf{B}}(t)$ as well as the $\underline{\mathbf{B}}_i$ being planes. The curve $\underline{\mathbf{B}}(t)$ is defined as the envelope of its *osculating planes* $\underline{\mathbf{B}}(t)$. This statement follows from the following comparison of properties of control point and control plane curves:

control point curve:		control plane curve:	
point:	$\underline{\mathbf{b}}(t)$	osculating plane:	$\underline{\mathbf{B}}(t)$
tangent:	$\underline{\mathbf{b}}(t) \wedge \dot{\underline{\mathbf{b}}}(t)$	tangent:	$\underline{\mathbf{B}}(t) \wedge \dot{\underline{\mathbf{B}}}(t)$
osculating plane:	$< \underline{\mathbf{b}}, \dot{\underline{\mathbf{b}}}, \ddot{\underline{\mathbf{b}}} >$	point:	$< \underline{\mathbf{B}}, \dot{\underline{\mathbf{B}}}, \ddot{\underline{\mathbf{B}}} >$

Here, we have used the notation that $< \underline{\mathbf{a}}, \underline{\mathbf{b}}, \underline{\mathbf{c}} >$ defines a plane and that $< \underline{\mathbf{A}}, \underline{\mathbf{B}}, \underline{\mathbf{C}} >$ defines a point; see Section 1.11 for details.

We may again ask: given a control point curve, how do we express it as a control plane curve? Taking the cubic case as before, we have

$$\begin{aligned}
\underline{\mathbf{B}}(t) &= < \underline{\mathbf{b}}_0^1(t), \underline{\mathbf{b}}_1^1(t), \underline{\mathbf{b}}_2^1(t) > \\
&= (1-t)^3 < \underline{\mathbf{b}}_0, \underline{\mathbf{b}}_1, \underline{\mathbf{b}}_2 > \\
&\quad + t(1-t)^2 < \underline{\mathbf{b}}_0, \underline{\mathbf{b}}_1, \underline{\mathbf{b}}_3 > \\
&\quad + t^2(1-t) < \underline{\mathbf{b}}_0, \underline{\mathbf{b}}_2, \underline{\mathbf{b}}_3 > \\
&\quad + t^3 < \underline{\mathbf{b}}_1, \underline{\mathbf{b}}_2, \underline{\mathbf{b}}_3 > .
\end{aligned}$$

In the plane case, the dual of a control point cubic is a control line quartic – yet for 3-D curves, the dual of a control point cubic is a control plane cubic! For higher degrees, however, we will see the degrees going up again: the dual of a degree n control point curve is a degree $3n-3$ control plane curve.

The importance of these dual developments is in the theory of developable surfaces, which we shall discuss in Section 11.8. The key observation will be that while the planes $\underline{\mathbf{B}}(t)$ are osculating planes of a curve, they themselves, being a one parameter family of planes, envelope a surface. Such surfaces are called developable.

7.12 Hybrid Bézier Curves

It is frequently important to approximate a rational Bézier curve by an integral, i.e., polynomial one. This need will arise when a rational Bézier curve is produced in one CAD system and is to be imported into another system, which can only handle integral curves. A direct conversion rational → polynomial is impossible, and so one must resort to approximation methods.

This section presents an approximation method due to Sederberg and Kakimoto [107], with an extension by Worsey, Sederberg, and Wang [119].

The basic idea is this: any rational Bézier curve

$$\mathbf{b}(t) = \frac{\sum_{i=0}^{n} w_i \mathbf{b}_i B_i^n(t)}{\sum_{i=0}^{n} w_i B_i^n(t)} \tag{7.43}$$

can be split into a polynomial and a rational part as follows. The polynomial part can be of arbitrary degree m. But one of the polynomial control points, $\mathbf{p}_s$, is given special treatment: the point $\mathbf{p}_s$ will not be constant; instead it will vary along a rational Bézier curve of the original degree n and with the original weights w_i. The hybrid form of a rational Bézier curve then takes the form

$$\mathbf{b}(t) = \sum_{\substack{i=0 \\ i \neq s}}^{m} \mathbf{p}_i B_i^m(t) + \frac{\sum_{i=0}^{n} w_i \mathbf{c}_i B_i^n(t)}{\sum_{i=0}^{n} w_i B_i^n(t)} \cdot B_s^m(t). \tag{7.44}$$

The numbers s and m are arbitrary — after they are selected, one has to determine the unknowns $\mathbf{p}_i$ and $\mathbf{c}_i$. We do this by equating the right-hand sides of (7.43) and (7.44):

$$\sum_{i=0}^{n} w_i \mathbf{b}_i B_i^n(t) = \Big(\sum_{i=0}^{n} w_i B_i^n(t)\Big)\Big(\sum_{\substack{i=0 \\ i \neq s}}^{m} \mathbf{p}_i B_i^m(t)\Big) + \Big(\sum_{i=0}^{n} w_i \mathbf{c}_i B_i^n(t)\Big) B_s^m(t). \tag{7.45}$$

The degree of the left-hand side curve can be elevated from n to $n+m$. We make use of the *generalized degree elevation formula:*

$$\sum_{i=0}^{n} \mathbf{b}_i B_i^n(t) = \sum_{i=0}^{n+m} \mathbf{b}_i^{(m)} B_i^{n+m}(t)$$

with

$$\mathbf{b}_i^{(m)} = \sum_{j=0}^{n} \mathbf{b}_j \binom{n}{j} \frac{\binom{m}{i-j}}{\binom{n+m}{i}} = \sum_{j+k=i} \mathbf{b}_i \binom{n}{j} \frac{\binom{m}{k}}{\binom{n+m}{i}}.$$

We will also use the product formula

$$B_i^n(t)B_j^m(t) = \frac{\binom{n}{i}\binom{m}{j}}{\binom{n+m}{i+j}} B_{i+j}^{n+m}(t),$$

thus transforming (7.45) into

$$\begin{aligned}
&\sum_{i=0}^{n+m} \Big(\sum_{j+k=i} w_j \mathbf{b}_j \frac{\binom{n}{j}\binom{m}{k}}{\binom{n+m}{i}} \Big) B_i^{n+m}(t) \\
&= \sum_{i=0}^{n+m} \Big(\sum_{j+k=i} w_j \mathbf{p}_k \frac{\binom{n}{j}\binom{m}{k}}{\binom{n+m}{i}} \Big) B_i^{n+m} + \sum_{i=s}^{s+n} w_{i-s}\mathbf{c}_{i-s} \frac{\binom{m}{s}\binom{n}{i-s}}{\binom{n+m}{i}} B_i^{n+m}(t).
\end{aligned}$$

For $i < s$, this simplifies enough so that we can compare coefficients:

$$\sum_{j+k=i} w_j \mathbf{b}_j \binom{n}{j}\binom{m}{k} = \sum_{j+k=i} w_j \mathbf{p}_k \binom{n}{j}\binom{m}{k}.$$

This allows us to compute the unknown $\mathbf{p}_i$ for $i = 0, \ldots, s-1$:

$$\mathbf{p}_0 = \mathbf{b}_0, \quad \mathbf{p}_1 = \mathbf{b}_0 + \frac{w_1 n}{w_0 m}\mathbf{b}_1 - \frac{w_1 n}{w_0 m}\mathbf{p}_0, \quad \text{etc.}$$

Analogously, we obtain $\mathbf{p}_m, \ldots, \mathbf{p}_{s+1}$ as

$$\mathbf{p}_m = \mathbf{b}_n, \quad \mathbf{p}_{m-1} = \mathbf{b}_n + \frac{w_{n-1} m}{w_n n}\mathbf{b}_{n-1} - \frac{w_{n-1} m}{w_n n}\mathbf{p}_m, \quad \text{etc.}$$

Having found all $\mathbf{p}_i$, we still need to determine the $\mathbf{c}_i$. For $i = s, \ldots, s+n$, we obtain

$$\sum_{j+k=i} w_j \mathbf{b}_j \binom{n}{j}\binom{m}{k} = \sum_{j+k=i} w_j \mathbf{p}_k \binom{n}{j}\binom{m}{k} + w_{i-s}\mathbf{c}_{i-s}\binom{m}{s}\binom{n}{i-s}$$

and thus

$$\mathbf{c}_{i-s} = \frac{\sum_{j+k=i} w_j \binom{n}{j}\binom{m}{k}[\mathbf{b}_j - \mathbf{p}_k]}{w_{i-s}\binom{m}{s}\binom{n}{i-s}}.$$

What is the benefit of this much algebra? Consider Figure 7.10. A rational quartic (control polygon not shown) is written as an integral quadratic, with the control point $\mathbf{b}_1$ replaced by another rational quartic. Notice that the size of this rational quartic is considerably less than that of the original curve. Loosely speaking, the size of the rational component tells us how "rational" the original curve was: the smaller the size of the rational component, the closer the original curve was to being polynomial.

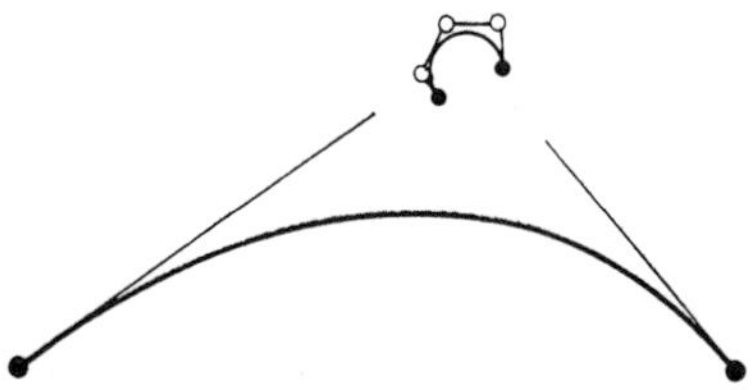

Figure 7.10. Hybrid curves: a rational quartic (control polygon not shown) is written as an integral quadratic with a variable center control point.

As we increase the degree m of the integral curve, the size of the rational component shrinks — as m tends to infinity, it will shrink to a point. This gives the desired algorithm for the approximation of a rational curve by a polynomial one: increase m until the size of the rational component (conveniently measured in terms of its minmax box) shrinks below a given tolerance, then replace the rational component by the center of its minmax box.

A. Worsey, T. Sederberg, and G. Wang [119] have streamlined the process of increasing the degree of the integral component. Instead of recomputing everything for each new value of m, one can capitalize on what has already been worked out. We discuss the case of a rational cubic, setting $m = 2$ and $s = 1$. We then have

$$\mathbf{b}(t) = \frac{\sum_{i=0}^{n} w_i \mathbf{b}_i B_i^n(t)}{\sum_{i=0}^{n} w_i B_i^n(t)} = \sum_{\substack{i=0\\ i\neq 1}}^{2} \mathbf{p}_i^{(1)} B_i^2(t) + \frac{\sum_{i=0}^{3} w_i \mathbf{c}_i^{(1)} B_i^3(t)}{\sum_{i=0}^{n} w_i B_i^3(t)} B_1^2(t),$$

from which we get

$$\begin{aligned}
\mathbf{p}_0^{(1)} &= \mathbf{b}_0, \\
\mathbf{p}_2^{(1)} &= \mathbf{b}_3, \\
\mathbf{c}_0^{(1)} &= \mathbf{b}_0 + \frac{3w_1}{2w_0}[\mathbf{b}_1 - \mathbf{b}_0], \\
\mathbf{c}_1^{(1)} &= \mathbf{b}_1 + \frac{1}{6w_1}[w_0(\mathbf{b}_0 - \mathbf{b}_3) + 3w_2(\mathbf{b}_2 - \mathbf{b}_0)], \\
\mathbf{c}_2^{(1)} &= \mathbf{b}_2 + \frac{1}{6w_2}[w_3(\mathbf{b}_3 - \mathbf{b}_0) + 3w_1(\mathbf{b}_1 - \mathbf{b}_3)], \\
\mathbf{c}_3^{(1)} &= \mathbf{b}_3 + \frac{3w_2}{2w_3}[\mathbf{b}_2 - \mathbf{b}_3].
\end{aligned}$$

The iteration now yields an expression for the rational component as yet another hybrid curve:

$$\mathbf{b}(t) = \sum_{\substack{i=0\\i\neq 1}}^{2} \mathbf{p}_i^{(1)} B_i^2(t) + \Big[\sum_{\substack{i=0\\i\neq 1}}^{2} \mathbf{p}_i^{(2)} B_i^2(t) \frac{\sum_{i=0}^{3} w_i \mathbf{c}_i^{(2)} B_i^3(t)}{\sum_{i=0}^{n} w_i B_i^3(t)}\Big] B_1^2(t). \qquad (7.46)$$

The recursion proceeds by continually replacing the rational component of each representation by another hybrid form. It stops when the size of a rational component's minmax box shrinks below a preset tolerance.

A different approach, based upon approximation, can be found in [12], [71], [72], [86].

7.13 Problems

1. Sketch (manually) the curve from Example 7.1, including the point $\mathbf{b}(2)$.

2. Experiment graphically with control line curves. Compute your curves in projective space, then project into affine space for display.

3. If we take successive derivatives of a polynomial curve, eventually they are all zero. What about successive derivatives of rational Bézier curves?

4. Equation (7.23) describes a hyperbola. It may be written in parametric form with control points $(0,0), (\alpha, \beta), (1,1)$. If in standard form, discuss how the middle weight and α, β depend on c.

5. Derive expressions for higher derivatives of a rational Bézier curve at $t = 0$, thus generalizing (7.29).

6. Experiment numerically with the two given versions of the de Casteljau algorithm: compare the completely rational version with the one that carries out all computations in projective space. In order to detect different behaviors, you will need to use rapidly varying weights.

8

Rational Cubics

Rational cubics are the simplest and most fundamental of all rational space curves. They are therefore important enough to warrant a whole chapter dedicated to their study alone. An excellent treatment of rational cubics is the one by Boehm [21]; another good reference is Forrest [59]. The more theoretically inclined reader will find valuable material in older books, such as [8] or [87].

8.1 Straight Line Cubics

If the four control points $\mathbf{b}_0, \mathbf{b}_1, \mathbf{b}_2, \mathbf{b}_3$ of a rational cubic

$$\mathbf{b}(t) = \frac{w_0\mathbf{b}_0 B_0^3(t) + w_1\mathbf{b}_1 B_1^3(t) + w_2\mathbf{b}_2 B_2^3(t) + w_3\mathbf{b}_3 B_3^3(t)}{w_0 B_0^3(t) + w_1 B_1^3(t) + w_2 B_2^3(t) + w_3 B_3^3(t)} \tag{8.1}$$

in affine space are spaced evenly on a straight line, and if the weights are unity, then the rational cubic is in fact polynomial, tracing out the straight line in a linear fashion. This fact is known as the *linear precision* property of Bézier curves. Note that $\mathbf{b}(\infty)$ is a point at infinity.

In the rational case, there are many ways of reproducing a straight line: suppose the control points are given by

$$\mathbf{b}_0, (1-\alpha)\mathbf{b}_0 + \alpha\mathbf{b}_3, \alpha\mathbf{b}_0 + (1-\alpha)\mathbf{b}_3, \mathbf{b}_3,$$

i.e., they are still distributed symmetrically, but not necessarily evenly. Assuming $w_0 = w_3 = 1$, can we find weights w_1 and w_2 such that $\mathbf{b}(t)$ traces out the line $\overline{\mathbf{b}_0, \mathbf{b}_3}$ linearly? By symmetry, we may assume $w_1 = w_2 = w$, which has the effect of making the denominator in fact quadratic.[1] We can determine the unknown w from

$$\frac{\mathbf{b}_0 B_0^3(t) + w\mathbf{b}_1 B_1^3(t) + w\mathbf{b}_2 B_2^3(t) + \mathbf{b}_3 B_3^3(t)}{B_0^3(t) + wB_1^3(t) + wB_2^3(t) + B_3^3(t)} = (1-t)\mathbf{b}_0 + t\mathbf{b}_3.$$

Since the denominator is actually quadratic, we may cross multiply and compare coefficients, arriving at

$$w = \frac{1}{3\alpha}. \tag{8.2}$$

An application is given by the following situation: suppose we are given a linear control polygon as described above, and we are asked to produce "good" weights. Using (8.2), we have likely found a good solution.

8.2 Rational Cubic Conics

While rational cubics are space curves in general, an important subclass is those rational cubics that are actually conics, and hence planar. In projective space, we may obtain a cubic from a quadratic (a conic) by the process of degree elevation.

In $\mathbb{P}^2$, consider Figure 8.1.[2] Let $\underline{\mathbf{b}}_0, \underline{\mathbf{b}}_1, \underline{\mathbf{b}}_2$ be the original control points of the conic, and let $\underline{\mathbf{c}}_0, \underline{\mathbf{c}}_1, \underline{\mathbf{c}}_2, \underline{\mathbf{c}}_3$ be the control points of the corresponding cubic. They are related by

$$\begin{aligned}
\underline{\mathbf{c}}_0 &= \underline{\mathbf{b}}_0, \\
\underline{\mathbf{c}}_1 &= \frac{1}{3}\underline{\mathbf{b}}_0 + \frac{2}{3}\underline{\mathbf{b}}_1, \\
\underline{\mathbf{c}}_2 &= \frac{2}{3}\underline{\mathbf{b}}_1 + \frac{1}{3}\underline{\mathbf{b}}_2, \\
\underline{\mathbf{c}}_3 &= \underline{\mathbf{b}}_2.
\end{aligned}$$

In affine space, the corresponding relationships are given by

$$\mathbf{c}_0 = \frac{\mathbf{b}_0}{z_0},$$

[1] Rational cubics with a quadratic denominator are called *T-conics* and will be discussed in Section 8.5.

[2] When dealing with conics, there is obviously no need to go to $\mathbb{P}^3$!

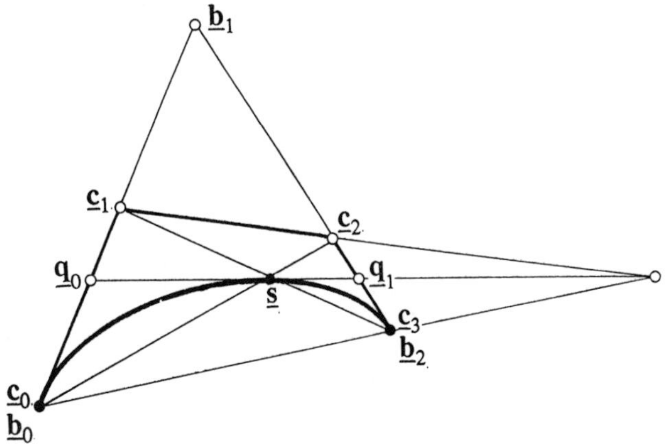

Figure 8.1. Cubic conics: the geometry of the quadratic → cubic degree elevation process in $\mathbb{P}^2$.

$$\begin{aligned}
\mathbf{c}_1 &= \frac{\mathbf{b}_0 + 2\mathbf{b}_1}{z_0 + 2z_1}, \\
\mathbf{c}_2 &= \frac{2\mathbf{b}_1 + \mathbf{b}_2}{2z_1 + z_2}, \\
\mathbf{c}_3 &= \frac{\mathbf{b}_2}{z_2},
\end{aligned}$$

spelling out the more general equation (8.8). The weights w_i of the cubic are then given by

$$\begin{aligned}
w_0 &= z_0, \\
w_1 &= z_0 + 2z_1, \\
w_2 &= 2z_1 + z_2, \\
w_3 &= z_2.
\end{aligned}$$

There is an interesting interplay between the quadratic and cubic forms. In projective space, we have

$$\underline{\mathbf{s}} = \underline{\mathbf{b}}(\frac{1}{2}) = \underline{\mathbf{c}}(\frac{1}{2}) = [\underline{\mathbf{c}}_0 \wedge \underline{\mathbf{c}}_2] \wedge [\underline{\mathbf{c}}_3 \wedge \underline{\mathbf{c}}_1]. \tag{8.3}$$

For a proof, we write

$$\underline{\mathbf{s}} = (1-\alpha)\underline{\mathbf{b}}_0 + \alpha(\frac{2}{3}\underline{\mathbf{b}}_1 + \frac{1}{3}\underline{\mathbf{b}}_2) = (1-\beta)\underline{\mathbf{b}}_2 + \beta(\frac{1}{3}\underline{\mathbf{b}}_0 + \frac{2}{3}\underline{\mathbf{b}}_1),$$

from which we deduce $\alpha = \beta = \frac{3}{4}$. Thus

$$\underline{\mathbf{s}} = \frac{1}{4}\underline{\mathbf{b}}_0 + \frac{1}{2}\underline{\mathbf{b}}_1 + \frac{1}{4}\underline{\mathbf{b}}_2 = \underline{\mathbf{c}}(\frac{1}{2}).$$

Let $\underline{\mathbf{q}}_0$ and $\underline{\mathbf{q}}_1$ be the intersections of the shoulder tangent $\underline{\mathbf{T}}(\frac{1}{2})$ with the quadratic control polygon, as shown in Figure 8.1. Since $\underline{\mathbf{q}}_0 = \frac{1}{2}\underline{\mathbf{b}}_0 + \frac{1}{2}\underline{\mathbf{b}}_1$ and $\underline{\mathbf{q}}_1 = \frac{1}{2}\underline{\mathbf{b}}_1 + \frac{1}{2}\underline{\mathbf{b}}_2$, we find that

$$\mathrm{cr}(\underline{\mathbf{b}}_0, \underline{\mathbf{q}}_0, \underline{\mathbf{c}}_1, \underline{\mathbf{b}}_1) = \mathrm{cr}(\underline{\mathbf{b}}_1, \underline{\mathbf{q}}_1, \underline{\mathbf{c}}_2, \underline{\mathbf{b}}_2) = \frac{1}{2},$$

i.e., each set of four points is *harmonic*. This relationship holds true, of course, for the corresponding affine points. We also note that the lines $\underline{\mathbf{c}}_1 \wedge \underline{\mathbf{c}}_2$, $\underline{\mathbf{q}}_0 \wedge \underline{\mathbf{q}}_1$, and $\underline{\mathbf{b}}_0 \wedge \underline{\mathbf{b}}_2$ are concurrent, all passing through the point $\underline{\mathbf{b}}_2 - \underline{\mathbf{b}}_0$.

Let us define $\underline{\mathbf{p}}_i = \frac{1}{2}\underline{\mathbf{c}}_i + \frac{1}{2}\underline{\mathbf{c}}_{i+1}$; $i = 0, 1, 2$. (These points will be mapped to the affine weight points of the rational cubic.) It is not hard to show that the three points $\underline{\mathbf{b}}_1, \underline{\mathbf{p}}_1$, and $\underline{\mathbf{s}}$ are collinear and that the line $\underline{\mathbf{p}}_0 \wedge \underline{\mathbf{p}}_2$ passes through the point $\underline{\mathbf{b}}_2 - \underline{\mathbf{b}}_0$.

We now consider the following affine problem: Given four points $\mathbf{c}_0$, $\mathbf{c}_1$, $\mathbf{c}_2$, $\mathbf{c}_3$ forming a convex and planar polygon, is it possible to find weights w_i such that the resulting rational cubic is actually a conic? The answer is "yes," and the weights can be found as follows:

First, we find the intersection $\mathbf{b}_1$ of the first and last polygon legs. Then $\mathbf{q}_0$, a conic weight point, is determined by the requirement that $\mathrm{cr}(\mathbf{c}_0, \mathbf{q}_0, \mathbf{c}_1, \mathbf{b}_1) = \frac{1}{2}$. Setting $z_0 = 1$, we then have

$$z_1 = \mathrm{ratio}(\mathbf{c}_0, \mathbf{q}_0, \mathbf{b}_1) = \frac{1}{2}\mathrm{ratio}(\mathbf{c}_0, \mathbf{c}_1, \mathbf{b}_1).$$

Here, z_1 is a conic weight. Proceeding in the same manner again, we find that

$$z_2 = z_1 \mathrm{ratio}(\mathbf{b}_1, \mathbf{q}_1, \mathbf{c}_3) = \frac{1}{2} z_1 \mathrm{ratio}(\underline{\mathbf{b}}_1, \mathbf{c}_2, \mathbf{c}_3).$$

The three weights $1, z_1, z_2$ are the weights of the conic; degree elevation now gives the desired cubic weights.

This method shows how we could estimate "good" weights for a rational cubic if only its control polygon were given and the weights had to be made up. The method does not work, understandably, for nonconvex or nonplanar polygons.

The technique may also be viewed as one for "degree reduction" for rational cubics with convex control polygons: if a rational cubic is given and a rational quadratic approximation is desired, its weights are z_0, z_1, z_2 and the control polygon is $\mathbf{c}_0, \mathbf{b}_1, \mathbf{c}_3$.

8.3 Embedding Cones

For every rational cubic, we can find a quadratic cone such that it "winds around" that cone. A quadratic cone is defined as follows: it is a surface with the three properties that

a) the tangent plane to any point of the cone has a whole line in common with the cone,

b) all those lines are concurrent,

c) any planar section of the cone is a conic, i.e, a quadratic curve.

Before we elaborate further, it should be noted that any space curve $\underline{\mathbf{x}}(t)$ lies on a cone: just pick any point $\underline{\mathbf{p}}$ and "cone off," i.e., form lines $\underline{\mathbf{p}} \wedge \underline{\mathbf{x}}(t)$. These lines envelop a conical surface. We will show here that for the special case of parametric cubics, we can find cones which are quadratic.

We will investigate this phenomenon in projective space $\mathbb{P}^3$. For a given cubic, define two *Ball points*:

$$\underline{\mathbf{l}}_1 = \frac{3}{2}\underline{\mathbf{b}}_1 - \frac{1}{2}\underline{\mathbf{b}}_0, \quad \underline{\mathbf{l}}_2 = \frac{3}{2}\underline{\mathbf{b}}_2 - \frac{1}{2}\underline{\mathbf{b}}_3, \tag{8.4}$$

after A. Ball, who introduced these points for use in his design system CONSURF;, see [9], [10], [11].

We may write our cubic using the Ball points:

$$\underline{\mathbf{b}}(t) = (1-t)^2\underline{\mathbf{b}}_0 + 2(1-t)^2 t\underline{\mathbf{l}}_1 + 2t^2(1-t)\underline{\mathbf{l}}_2 + t^2\underline{\mathbf{b}}_3. \tag{8.5}$$

This is proved by simply inserting $\underline{\mathbf{l}}_i$'s definition into the cubic Bézier form. In the special case that both $\underline{\mathbf{l}}_i$ agree: $\underline{\mathbf{l}} = \underline{\mathbf{l}}_1 = \underline{\mathbf{l}}_2$, our cubic is actually a quadratic, with control polygon $\underline{\mathbf{b}}_0, \underline{\mathbf{l}}, \underline{\mathbf{b}}_2$.

Assuming from now on that $\underline{\mathbf{l}}_1 \neq \underline{\mathbf{l}}_2$, let $\underline{\mathbf{P}}$ be any plane not containing the *Ball line* $\underline{\mathbf{L}} = \underline{\mathbf{l}}_1 \wedge \underline{\mathbf{l}}_2$ and let $\underline{\mathbf{v}}$ be a point on that line. Let us now project our cubic into $\underline{\mathbf{P}}$ through $\underline{\mathbf{v}}$. Then the images of the two Ball points coincide, yielding a planar image control polygon $\hat{\underline{\mathbf{b}}}_0, \hat{\underline{\mathbf{l}}}, \hat{\underline{\mathbf{b}}}_3$; see Figure 8.2. It defines a conic $\underline{\mathbf{c}}(t)$ in the plane $\underline{\mathbf{P}}$ which is the projection of the original cubic. As every line $\underline{\mathbf{c}}(t) \wedge \underline{\mathbf{b}}(t)$ passes through $\underline{\mathbf{v}}$, we find that the curve $\underline{\mathbf{b}}(t)$ is on a quadratic cone with generating lines $\underline{\mathbf{c}}(t) \wedge \underline{\mathbf{b}}(t)$ and with vertex $\underline{\mathbf{v}}$. Note that the Ball line $\underline{\mathbf{L}}$ itself does not lie on the cone; it intersects it at the point $\underline{\mathbf{v}}$, the cone's vertex.

Different choices of $\underline{\mathbf{v}}$ on $\underline{\mathbf{L}}$ will give rise to different cones, and so there is in fact a one-parameter family of cones on which $\underline{\mathbf{b}}(t)$ lies. Of all points on $\underline{\mathbf{L}}$, one is special with respect to the given cubic: it is the difference of

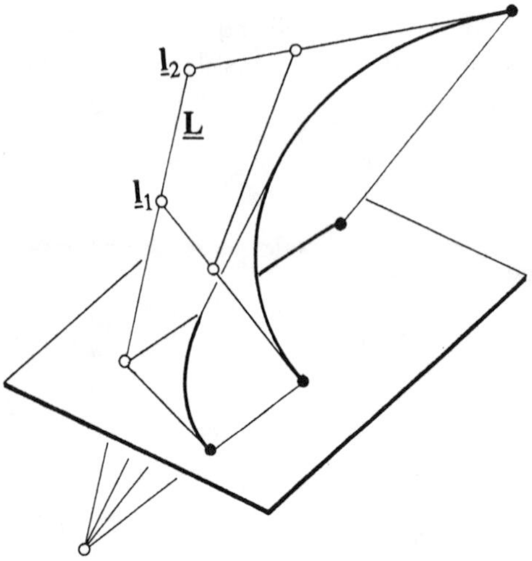

Figure 8.2. Projecting a cubic: a cubic is projected into a plane such that the two Ball points have identical images.

the Ball points:

$$\begin{aligned}\underline{\mathbf{l}}_1 - \underline{\mathbf{l}}_2 &= \frac{1}{2}(3\underline{\mathbf{b}}_1 - \underline{\mathbf{b}}_0 - 3\underline{\mathbf{b}}_2 + \underline{\mathbf{b}}_3) \\ &= \frac{1}{2}\Delta^3 \underline{\mathbf{b}}_0 \\ &\hat{=} \ \underline{\mathbf{b}}(\infty) \\ &\hat{=} \ \dot{\underline{\mathbf{b}}}(\infty) \\ &\hat{=} \ \ddot{\underline{\mathbf{b}}}(\infty).\end{aligned}$$

Thus the point $\underline{\mathbf{b}}(\infty)$ lies on $\underline{\mathbf{L}}$.

If we reparametrize the cubic using a Moebius transformation, the Ball points will change; but again $\underline{\mathbf{l}}_1 - \underline{\mathbf{l}}_2$ will be on the cubic! Figure 8.3 shows how the Ball line is affected by reparametrizations.

As we project our cubic into affine space, we obtain a rational cubic with Ball points

$$\mathbf{l}_1 = \frac{3w_1\mathbf{b}_1 - w_0\mathbf{b}_0}{3w_1 - w_0}, \mathbf{l}_2 = \frac{3w_2\mathbf{b}_2 - w_3\mathbf{b}_3}{3w_2 - w_3}. \tag{8.6}$$

As cones are mapped to cones by projections, the rational cubic also lies on a one-parameter family of cones. The Ball line $\mathbf{L}$ again intersects the

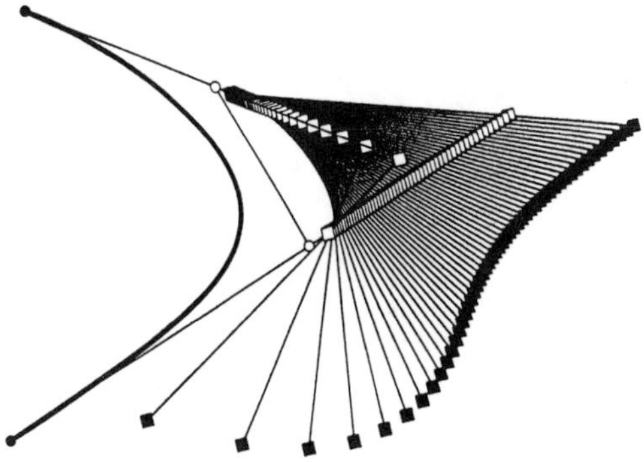

Figure 8.3. Ball lines: depending on its parametrization, a cubic possesses different Ball lines. Ball points are marked by hollow squares, their differences are marked by solid squares.

rational cubic for $t = \infty$. In the case of an integral cubic, the cone becomes a *parabolic cylinder*, and $\mathbf{L}$ is called an *asymptotic direction*, meaning that for $t \to \infty$, the curve will have tangents parallel to $\mathbf{L}$. This is seen easily by observing that the dominating term in $\lim_{t\to\infty} \dot{\mathbf{b}}(t)$ is given by $\Delta^3 \mathbf{b}_0$ wich equals $\mathbf{l}_2 - \mathbf{l}_1$. Figure 8.4 illustrates the integral case.

8.4 The Embedding Bilinear Patch

Let a projective cubic Bézier curve be given by its control points $\underline{\mathbf{b}}_0, \underline{\mathbf{b}}_1, \underline{\mathbf{b}}_2, \underline{\mathbf{b}}_3$; we define

$$\underline{\mathbf{a}}_1 = \frac{3}{2}\underline{\mathbf{b}}_1 - \frac{1}{2}\underline{\mathbf{b}}_3, \quad \underline{\mathbf{a}}_2 = \frac{3}{2}\underline{\mathbf{b}}_2 - \frac{1}{2}\underline{\mathbf{b}}_0. \tag{8.7}$$

Then the four points $\underline{\mathbf{b}}_0, \underline{\mathbf{a}}_1, \underline{\mathbf{a}}_2, \underline{\mathbf{b}}_3$ form a bilinear surface (see Section 11.1 for details):

$$\underline{\mathbf{x}}(u, v) = \underline{\mathbf{b}}_0(1-u)(1-v) + \underline{\mathbf{b}}_3 u(1-v) + \underline{\mathbf{a}}_1(1-u)v + \underline{\mathbf{a}}_2 uv. \tag{8.8}$$

This surface is a doubly ruled quadric, a so-called annular quadric. To show that our cubic lies on this surface, we set

$$v = B_1^2(u) = 2(1-u)u$$

and insert this into (8.8): we recover the equation of the cubic! See Figure 8.5 for an illustration.

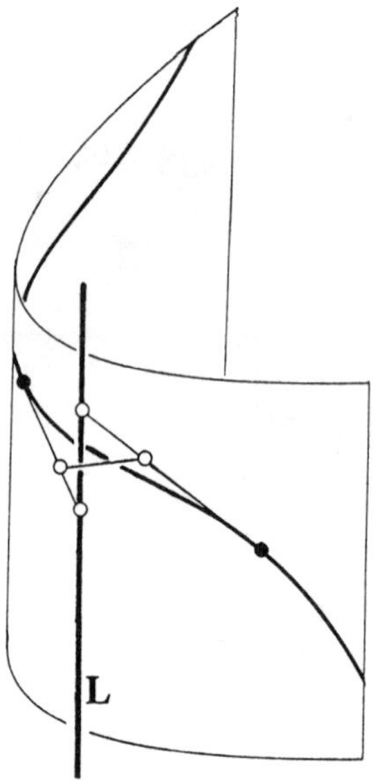

Figure 8.4. Integral cubics: the curve lies on a parabolic cylinder. The cylinder axis is parallel to the Ball line.

After we project (8.8) into affine space, the projective bilinear surface becomes an affine rational bilinear patch. The four corners of this patch are given by the projections $\mathbf{b}_i$ and $\mathbf{a}_i$ of the $\underline{\mathbf{b}}_i$ and the $\underline{\mathbf{a}}_i$. We note that

$$\mathbf{a}_1 = \frac{3w_1\mathbf{b}_1 - w_3\mathbf{b}_3}{3w_1 - w_3} \quad \text{and} \quad \mathbf{a}_2 = \frac{3w_2\mathbf{b}_2 - w_0\mathbf{b}_0}{3w_2 - w_0}.$$

Together with the result from Section 8.3, we see that every rational cubic is the intersection between a cone and a bilinear patch. For the integral case, when all four w_i are equal, the embedding bilinear patch is a *hyperbolic paraboloid.*

8.5 Asymptotes

A cubic[3] $\underline{\mathbf{b}}(t)$ in $\mathbb{P}^3$ intersects every plane at least once. In particular, it intersects the plane $\underline{\mathbf{H}}$ which will be mapped to the plane at infinity when

[3]Here, we refer to actual cubics, not quadratics who were made cubic by degree elevation.

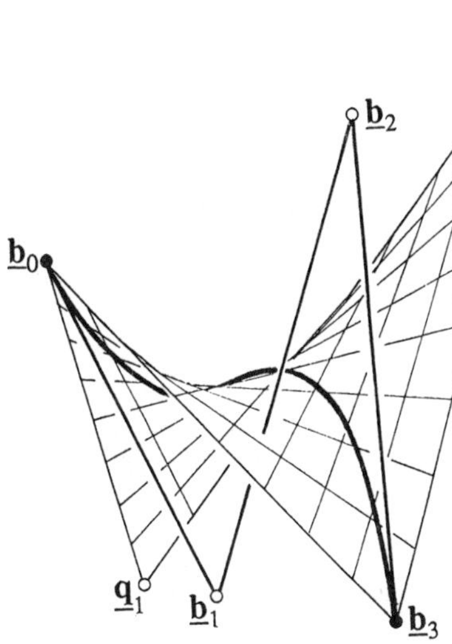

Figure 8.5. The embedding bilinear patch: every cubic lies entirely on a doubly ruled quadric.

we project into affine space. Let $\underline{\mathbf{x}}$ be such a point of intersection; also let $\underline{\mathbf{T}}$ be the cubic's tangent at $\underline{\mathbf{x}}$. As we project into affine space, $\underline{\mathbf{x}}$ will be mapped to a point at infinity, and $\underline{\mathbf{T}}$ will be mapped to an *asymptote*. An asymptote is thus the curve's tangent at infinity. The reader is invited to compare this situation with the similar case of hyperbolas; see Figure 3.15.

What happens when $\underline{\mathbf{H}}$ is *tangent* to the cubic? Then the resulting (affine) asymptote is itself at infinity, and the rational cubic touches the plane at infinity. Note that there are two different cases: the projective cubic may touch $\underline{\mathbf{H}}$ (just as $y = (1-x)^3$ touches the $x-$axis at $x = 1$) or it may intersect $\underline{\mathbf{H}}$ (just as $y = x^3$ intersects the $x-$axis at $x = 0$).

Algebraically speaking, the presence of an asymptote means that the denominator $w(t)$ of the rational cubic $\mathbf{b}(t)$ has at least one real zero t_0; as $t \to t_0$, the curve $\mathbf{b}(t)$ tends to infinity. Fig. 8.6 illustrates the geometry.

We could now reparametrize our rational cubic such that we map $0, 1, t_0$ to $0, 1, \infty$. Then the curve would tend to infinity as the new parameter s tends to infinity, thus resembling the behavior of a parabola in standard form (Section 5.2). Since

$$\lim_{s \to \infty} \mathbf{b}(s) = \frac{\Delta^3[\hat{w}_0 \mathbf{b}_0]}{\Delta^3 \hat{w}_0},$$

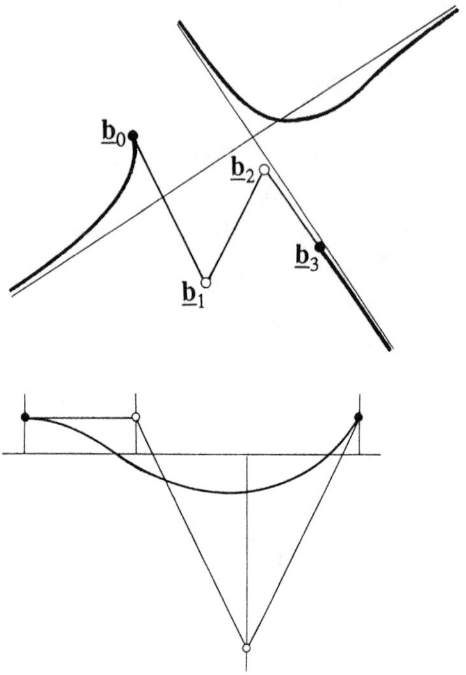

Figure 8.6. Asymptotes: the denominator (below) vanishes twice — the curve has two asymptotes (above).

it follows that $\Delta^3\hat{w}_0 = 0$, meaning that the new denominator is actually quadratic. Recalling that $\hat{w}_i = c^i w_i$ for some constant c, we can find c from

$$\Delta^3\hat{w}_0 = w_0 - 3cw_1 + 3c^2w_2 - c^3w_3 = 0.$$

With this new set of weights, the denominator can be written as a quadratic, and our rational cubic becomes

$$\mathbf{b}(s) = \frac{\sum_{i=0}^{3} \hat{w}_i \mathbf{b}_i B_i^3(s)}{\sum_{i=0}^{2} v_i B_i^2(s)}, \tag{8.9}$$

where the v_i are obtained from degree reducing the denominator. This form of a rational cubic was first considered by L. Cremona in 1859. It was named *T-conic* by M. Rowin [100] in 1964, as quoted from R. Forrest [57].

Writing a rational cubic in T-conic form simplifies the form of the asymptotes. We know that one asymptote (or asymptotic direction) is given by the Ball line $\mathbf{L}$.[4] There may be (at most) two more, and they are then given by the zeroes of the numerator $\sum_{i=0}^{2} v_i B_i^2(s)$. It would be useful to detect if these zeroes are in the interval $[0,1]$. No such zero exists if all three v_i are positive. If v_0 and v_2 are positive, then the condition $v_1 > -\sqrt{v_0 v_2}$ ensures that no zero is in $[0,1]$ (see Boehm [21]).

8.6 The Hermite Form

In affine geometry, *Hermite interpolation* is a well-known way to interpolate to curves when derivative information is given. See [34] or [47] for details. Here, we shall derive this curve form from a projective formulation.

Starting with a projective cubic Bézier curve, let us introduce $\underline{\mathbf{t}}_0 = \underline{\mathbf{b}}_1 - \underline{\mathbf{b}}_0$ and $\underline{\mathbf{t}}_1 = \underline{\mathbf{b}}_3 - \underline{\mathbf{b}}_2$. We can then write the original cubic as

$$\underline{\mathbf{b}}(t) = \underline{\mathbf{b}}_0 B_0^3(t) + [\underline{\mathbf{t}}_0 + \underline{\mathbf{b}}_0] B_1^3(t) + [\underline{\mathbf{b}}_3 - \underline{\mathbf{t}}_1] B_2^3(t) + \underline{\mathbf{b}}_3 B_3^3(t).$$

This may be reformulated as

$$\underline{\mathbf{b}}(t) = \underline{\mathbf{b}}_0 H_0^3(t) + 3\underline{\mathbf{t}}_0 H_1^3(t) + 3\underline{\mathbf{t}}_1 H_2^3(t) + \underline{\mathbf{b}}_3 H_3^3(t)$$

by setting

$$H_0^3 = B_0^3 + B_1^3, H_1^3 = \frac{1}{3} B_1^3, H_2^3 = -\frac{1}{3} B_2^3, H_3^3 = B_2^3 + B_3^3.$$

Note that $\dot{\underline{\mathbf{b}}}(0) \hat{=} \underline{\mathbf{t}}_0$ and $\dot{\underline{\mathbf{b}}}(1) \hat{=} \underline{\mathbf{t}}_1$ – but don't forget that $\dot{\underline{\mathbf{b}}}(0)$ and $\dot{\underline{\mathbf{b}}}(1)$ are points in $\mathbb{P}^3$!

Assuming that $\underline{\mathbf{b}}(t)$ is in standard form, i.e., $w_0 = w_3 = 1$, we project into affine space to obtain the rational form

$$\mathbf{b}(t) = \frac{\mathbf{b}_0 H_0^3(t) + 3v_0 \mathbf{t}_0 H_1^3(t) + 3v_1 \mathbf{t}_1 H_2^3(t) + \mathbf{b}_3 H_3^3(t)}{H_0^3(t) + 3v_0 H_1^3(t) + 3v_1 H_2^3(t) + H_3^3(t)}, \tag{8.10}$$

with v_0 and v_1 as the fourth coordinates of $\underline{\mathbf{t}}_0$ and $\underline{\mathbf{t}}_1$, respectively.

In affine space, derivatives correspond to vectors. If we want the projections of $\underline{\mathbf{t}}_0$ and $\underline{\mathbf{t}}_1$ to be derivatives of the rational curve $\mathbf{b}(t)$, then we must arrange for the projection to satisfy a special condition: it must map the line $\underline{\mathbf{H}} = \underline{\mathbf{t}}_0 \wedge \underline{\mathbf{t}}_1$ to an affine line at infinity. Then $v_0 = v_1 = 0$, and the $\underline{\mathbf{t}}_i$ are mapped to *tangent vectors* $\mathbf{t}_i$.[5] Since $H_0^3(t) + H_3^3(t) \equiv 1$, the rational curve (8.10) now takes on the simple polynomial form

$$\mathbf{b}(t) = \mathbf{b}_0 H_0^3(t) + 3\mathbf{t}_0 H_1^3(t) + 3\mathbf{t}_1 H_2^3(t) + \mathbf{b}_3 H_3^3(t), \tag{8.11}$$

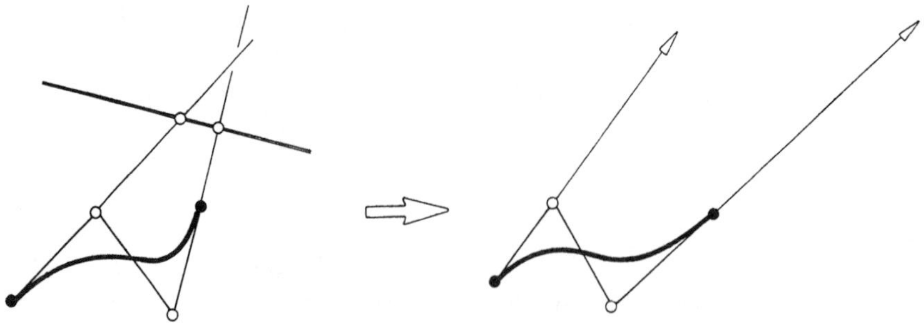

Figure 8.7. The Hermite form: left, as a polynomial in projective space, right: as a polynomial in affine space.

which is the familiar cubic Hermite interpolant. We easily verify that $\dot{\mathbf{b}}(0) = \mathbf{t}_0$ and $\dot{\mathbf{b}}(1) = \mathbf{t}_1$. Figure 8.7 gives an illustration.

We see that the Bézier form is taken easily from projective space, where it is just a polynomial curve, to affine space, where it is rational polynomial. This is because all of its coefficients are *points*. The Hermite form, with its affine point/vector mix of coefficients, does not allow allow this simple treatment.

8.7 Cubic Lagrange Interpolation

In Section 4.7, we interpolated to four coplanar points and corresponding parameter values. With cubics, we can move to projective three-space and interpolate to five points with corresponding parameter values. Thus we are given five points $\underline{\mathbf{x}}_0, \ldots, \underline{\mathbf{x}}_4$ and corresponding parameter values $t_0, \ldots, t_4$. We want a cubic $\underline{\mathbf{c}}(t)$ such that $\underline{\mathbf{c}}(t_i) \hat{=} \underline{\mathbf{x}}_i$.

We closely follow Section 4.7, and only give the main steps here. Accordingly, we write a point on the cubic as

$$\underline{\mathbf{c}}(t) = [B_0^3(t), B_1^3(t), B_2^3(t), B_3^3(t)] \begin{bmatrix} \underline{\mathbf{b}}_0 \\ \underline{\mathbf{b}}_1 \\ \underline{\mathbf{b}}_2 \\ \underline{\mathbf{b}}_3 \end{bmatrix}. \tag{8.12}$$

We will first consider the four interpolation conditions $w_i\underline{\mathbf{x}}_i = \underline{\mathbf{c}}(t_i); i =$

[4]The Ball line changes as we reprametrize to get to the T-conic!

[5]Compare with the control vector approach from Section 7.10!

0, 1, 2, 3, with unknown constants w_i. This leads to a 4×4 linear system

$$\boxed{\mathbf{x}} = B\boxed{\mathbf{b}}$$

with coefficient matrix $B = \{B_i^3(t_j)\}_{i,j=0}^3$ and solution $\boxed{\mathbf{b}} = B^{-1}\boxed{\mathbf{x}}$. The vector $\boxed{\mathbf{x}}$ still contains the unknowns w_i. Bringing in the fifth point $\underline{\mathbf{x}}_4$, we obtain $\underline{\mathbf{x}}_4 = B_4\boxed{\mathbf{b}}$ with $B_4 = [B_0^3(t_4), B_1^3(t_4), B_2^3(t_4), B_3^3(t_4)]$. Thus we obtain

$$\underline{\mathbf{x}}_4 = B_4 B^{-1}\boxed{\mathbf{x}},$$

which is a 4×4 linear system for the unknowns w_i. Once we solve for them, there is the possibility that some of them may be negative!

8.8 Osculatory Interpolation

With rational cubics, it is easy to solve an interesting kind of interpolation problem: given a Bézier polygon $\mathbf{b}_0, \mathbf{b}_1, \mathbf{b}_2, \mathbf{b}_3$ and a curvature value at each endpoint, find a set of weights w_0, w_1, w_2, w_3 such that the corresponding rational cubic assumes the given curvatures at $\mathbf{b}_0$ and $\mathbf{b}_1$. The following method is very similar to one developed by T. Goodman in 1988; see [64]. We assume without loss of generality that $w_0 = w_3 = 1$.[6] The given curvatures κ_0 and κ_3 are given by

$$\kappa_0 = \frac{4}{3}\frac{w_2}{w_1^2}c_0, \qquad \kappa_3 = \frac{4}{3}\frac{w_1}{w_2^2}c_1, \tag{8.13}$$

where

$$c_0 = \frac{\text{area}[\mathbf{b}_0, \mathbf{b}_1, \mathbf{b}_2]}{\text{dist}^3[\mathbf{b}_0, \mathbf{b}_1]}, \qquad c_1 = \frac{\text{area}[\mathbf{b}_1, \mathbf{b}_2, \mathbf{b}_3]}{\text{dist}^3[\mathbf{b}_2, \mathbf{b}_3]}.$$

Equations (8.13) decouple nicely, so that we can determine our unknowns w_1 and w_2:

$$w_1 = \frac{4}{3}\left[\frac{c_0^2}{\kappa_0^2}\frac{c_1}{\kappa_1}\right]^{\frac{1}{3}}, \qquad w_2 = \frac{4}{3}\left[\frac{c_0}{\kappa_0}\frac{c_1^2}{\kappa_1^2}\right]^{\frac{1}{3}}. \tag{8.14}$$

For planar control polygons, the quantities c_0 or c_1 may be negative – this happens when a control polygon is S-shaped. This is meaningful since curvature may be defined as *signed curvature* for 2-D curves. Of course, one should then also prescribe the corresponding κ_0 and κ_1 as being negative, so that one ends up with positive weights.

A similar interpolation problem was addressed by Klass [74] and de Boor, Hollig, and Sabin for the nonrational case: they prescribe two points and corresponding tangent directions and curvatures [37]. The solution (when it exists) can be obtained only using an iterative method.

---

[6] Goodman [64] assumes that $w_1 = w_2 = 1$.

8.9 Problems

1. In Section 8.4, we remarked that every rational cubic is the intersection between a cone and a bilinear patch. How about the converse: is every intersection between a cone and a bilinear patch a rational cubic? (This question leads into the area of algebraic geometry. Pointers to the literature: [1], [2], [3], [4].)

2. If a rational cubic is *closed*, does this mean that it is a conic?

3. Does the approach from Section 8.1 still work if $\mathbf{b}_1$ and $\mathbf{b}_2$ are not symmetrically distributed?

4. What is the model of a cubic in the spherical or circular models of the projective plane? If you have access to the appropriate kind of graphics equipment, illustrate!

5. Show that there are no closed nonplanar rational cubics.

9

Projective Splines

The most frequently used kind of spline curves are cubic B-splines. With the advent of NURBS, it is not very hard to predict that rational cubic B-splines will play a role of equal weight: rational cubic B-splines are a superset of both conic splines and integral cubic splines. Our development of rational cubic splines will be a purely projective one – in projective space, we are only dealing with piecewise polynomials, making smoothness questions a lot easier to handle.

9.1 Osculants and Smoothness

In affine space, it is easy to define second order continuity between two adjacent curve segments: just require that the corresponding second derivatives are continuous. In projective space, we are – at first sight – faced with a dilemma: continuity relies on the concept of distance,[1] but we have not even defined distance in projective space. Instead of embarking on this arduous path, we employ a concept that is tailored for our purposes: the osculants from Section 7.9; see Figure 7.9.

We will use osculants because they have a very useful property in pro-

[1]A curve is continuous if a small distance of two domain points implies a small distance of the corresponding curve points.

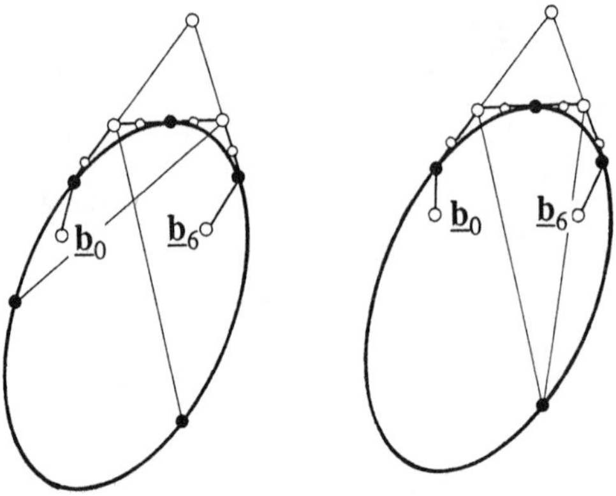

Figure 9.1. Smoothness: two G^2 cubics that share the same osculant. Left: G^2; right: C^2.

jective space: their derivatives agree with that of the cubic:[2]

$$\dot{\underline{\mathbf{x}}}(0) \hat{=} \dot{\underline{\mathbf{o}}}_0(0) \quad \text{and} \quad \ddot{\underline{\mathbf{x}}}(0) \hat{=} \ddot{\underline{\mathbf{o}}}_0(0). \tag{9.1}$$

A similar statement holds for $t = 1$. So instead of trying to define second order smoothness between two adjacent cubics, we will simply require that their osculants agree. This approach has also been taken by Pottmann [94] and Geise/Jüttler [61].

We may proceed in two ways: we may require that the osculants $\underline{\mathbf{o}}_-$ and $\underline{\mathbf{o}}_+$ agree as point sets, or we may require that the osculants are obtained from one conic by the process of subdivision, i.e., they are parametrized the same way. In the first case, we shall say that our two cubics join with G^2 continuity (G for geometric); in the second case, we shall define them to be C^2. Note that for C^2 continuity, we have that $\underline{\mathbf{o}}_-(\infty) = \underline{\mathbf{o}}_+(\infty)$. Figure 9.1 illustrates. To begin, we shall restrict ourselves to the case of G^2 continuity.

Now let $\underline{\mathbf{b}}_0, \ldots, \underline{\mathbf{b}}_3$ and $\underline{\mathbf{b}}_3, \ldots, \underline{\mathbf{b}}_6$, together with points $\underline{\mathbf{q}}_i$ on each polygon leg, define two cubics in projective space. If they share a common osculant at $\underline{\mathbf{b}}_3$, it is formed by $\underline{\mathbf{b}}_1, \underline{\mathbf{q}}_1, \underline{\mathbf{b}}_2, \underline{\mathbf{q}}_2, \underline{\mathbf{b}}_3$ and also by $\underline{\mathbf{b}}_3, \underline{\mathbf{q}}_3, \underline{\mathbf{b}}_4, \underline{\mathbf{q}}_4, \underline{\mathbf{b}}_5$. If these two conics form *one* conic $\underline{\mathbf{c}}(t)$, then $\underline{\mathbf{c}}$ may be defined by a control polygon $\underline{\mathbf{b}}_1, \underline{\mathbf{d}}, \underline{\mathbf{b}}_5$ and shoulder tangent $\underline{\mathbf{b}}_2 \wedge \underline{\mathbf{b}}_4$, where $\underline{\mathbf{d}}$ is the intersection of the osculant's tangents at $\underline{\mathbf{b}}_1$ and $\underline{\mathbf{b}}_5$. Figure 9.2 illustrates.

As a consequence, we note that

$$[\underline{\mathbf{b}}_1 \wedge \underline{\mathbf{b}}_4] \wedge [\underline{\mathbf{b}}_2 \wedge \underline{\mathbf{b}}_5], \underline{\mathbf{b}}_3, \underline{\mathbf{d}}$$

[2] Recall that this is a purely projective statement. It does not hold for rational cubics in affine space.

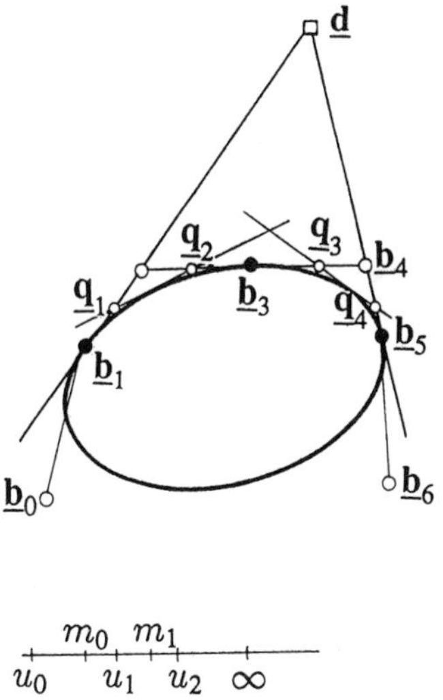

Figure 9.2. Osculants: if two osculants agree, they can be written as one conic.

are collinear, following the development in Section 3.4. We can thus compute $\underline{\mathbf{b}}_3$ from $\underline{\mathbf{b}}_1, \underline{\mathbf{b}}_2, \underline{\mathbf{b}}_4, \underline{\mathbf{b}}_5$.

From the four tangent theorem (see Section 3.3), it follows that several cross ratios must be equal; we list the following:

$$\operatorname{cr}(\underline{\mathbf{b}}_1, \underline{\mathbf{q}}_1, \underline{\mathbf{b}}_2, \underline{\mathbf{d}}) = \operatorname{cr}(\underline{\mathbf{b}}_2, \underline{\mathbf{q}}_2, \underline{\mathbf{b}}_3, \underline{\mathbf{b}}_4). \tag{9.2}$$

This allows us to determine $\underline{\mathbf{q}}_2$ if all $\underline{\mathbf{b}}_i$ and $\underline{\mathbf{q}}_1$ are given. Similarly, we may determine $\underline{\mathbf{q}}_3$ from

$$\operatorname{cr}(\underline{\mathbf{b}}_5, \underline{\mathbf{q}}_4, \underline{\mathbf{b}}_4, \underline{\mathbf{d}}) = \operatorname{cr}(\underline{\mathbf{b}}_4, \underline{\mathbf{q}}_3, \underline{\mathbf{b}}_3, \underline{\mathbf{b}}_2) \tag{9.3}$$

once $\underline{\mathbf{q}}_4$ is given.

To summarize: we can construct a G^2 projective spline with two cubic pieces from

a) a control polygon $\underline{\mathbf{b}}_0, \underline{\mathbf{b}}_1, \underline{\mathbf{d}}, \underline{\mathbf{b}}_5, \underline{\mathbf{b}}_6$,

b) interior Bézier points $\underline{\mathbf{b}}_2, \underline{\mathbf{b}}_4$,

c) weight points $\underline{\mathbf{q}}_0, \underline{\mathbf{q}}_1, \underline{\mathbf{q}}_4, \underline{\mathbf{q}}_5$.

Figure 9.3 gives an illustration of the closed case: appropriate selection of the interior Bézier points gives rise to different curve shapes.

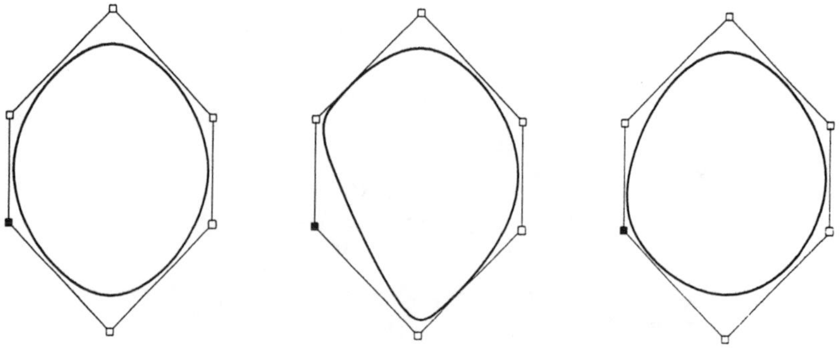

Figure 9.3. Closed splines: by an appropriate choice of weight points near the solid control point, different curve shapes are achieved.

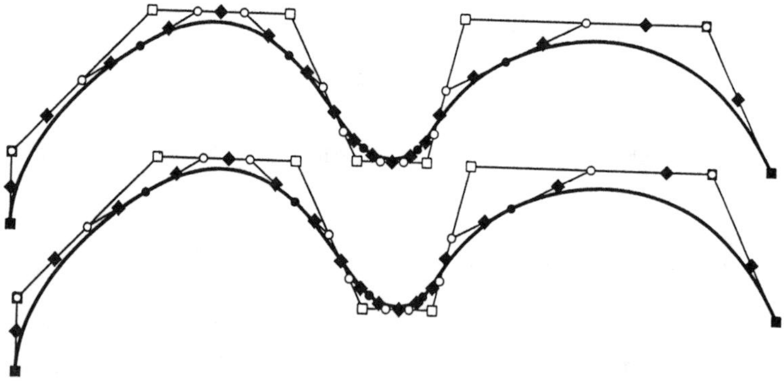

Figure 9.4. Reparametrization: top, an initial curve; bottom, after a reparametrization. Squares denote B-spline control vertices, circles denote Bézier control vertices, and diamonds denote the weight points.

9.2 Reparametrization

Upon close inspection, we can see that the specification of $\underline{\mathbf{q}}_1$ and $\underline{\mathbf{q}}_4$ has no influence on the shape of the resulting curve: if we change these two weight points, each cubic piece will be *reparametrized*; see Figure 9.4. This is a direct consequence of the four tangent theorem.

It is now natural to ask: can we reparametrize two G^2 cubics such that they become C^2? We assume that the two cubics are given by their respective control points $\underline{\mathbf{b}}_i$ and their weight points $\underline{\mathbf{q}}_i$. This question is still incomplete: we also have to specify the knot sequence with respect to which we want a C^2 curve. Our spline is fully defined once we have specified all $\hat{\underline{\mathbf{q}}}_i$ and the knot sequence u_i. In each interval, we will define $m_i = \frac{1}{2}u_i + \frac{1}{2}u_{i+1}$, in analogy to the development in Section 6.1. We construct the u_i as follows: Referring to Figure 9.2,

a) determine u_0, m_0, u_1 arbitrarily. Also determine $\hat{\underline{\mathbf{q}}}_0$ arbitrarily.

b) Determine $\hat{\underline{\mathbf{q}}}_1$ from

$$\mathrm{cr}(\underline{\mathbf{b}}_0, \underline{\mathbf{q}}_0, \hat{\underline{\mathbf{q}}}_0, \underline{\mathbf{b}}_1) = \mathrm{cr}(\underline{\mathbf{b}}_1, \underline{\mathbf{q}}_1, \hat{\underline{\mathbf{q}}}_1, \underline{\mathbf{b}}_2).$$

c) Find $\underline{\mathbf{q}}_2$ from (9.2).

d) Find u_2 from

$$\mathrm{cr}(u_0, m_0, u_1, u_2) = \mathrm{cr}(\underline{\mathbf{b}}_2, \underline{\mathbf{q}}_2, \underline{\mathbf{b}}_3, \underline{\mathbf{b}}_4).$$

e) Find m_1 from

$$\mathrm{cr}(u_0, m_0, u_1, \infty) = \mathrm{cr}(u_1, m_1, u_2, \infty).$$

f) Find $\underline{\mathbf{q}}_3$ and $\underline{\mathbf{q}}_4$ from

$$\mathrm{cr}(\underline{\mathbf{b}}_2, \underline{\mathbf{b}}_3, \underline{\mathbf{q}}_3, \underline{\mathbf{b}}_4) = \mathrm{cr}(\underline{\mathbf{d}}, \underline{\mathbf{b}}_4, \underline{\mathbf{q}}_4, \underline{\mathbf{b}}_5) = \mathrm{cr}(u_0, u_1, m_1, u_2).$$

g) Find $\hat{\underline{\mathbf{q}}}_5$ from

$$\mathrm{cr}(\underline{\mathbf{b}}_5, \underline{\mathbf{q}}_5, \hat{\underline{\mathbf{q}}}_5, \underline{\mathbf{b}}_6) = \mathrm{cr}(\underline{\mathbf{b}}_4, \underline{\mathbf{q}}_4, \hat{\underline{\mathbf{q}}}_4, \underline{\mathbf{b}}_5).$$

In the above construction, we picked u_0, m_0, u_1 arbitrarily. The question then arises: will a different choice give rise to a different spline curve? To keep matters simple, assume that we selected a different value for m_0, and name it $\hat{m}_0$. We may view this as a reparametrization of the domain, namely by a map $u_0, m_0, u_1 \rightarrow u_0, \hat{m}_0, u_1$. Such a map is, of course, a Moebius transformation. We would then simply determine $\hat{\underline{\mathbf{q}}}_0$ from

$$\mathrm{cr}(u_0, m_0, \hat{m}_0, u_1) = \mathrm{cr}(\underline{\mathbf{b}}_0, \underline{\mathbf{q}}_0, \hat{\underline{\mathbf{q}}}_0, \underline{\mathbf{b}}_1)$$

and proceed as above. The next section gives an illustration.

9.3 Projective Cubic Splines

Of course, we can form curves that consist of more than just two pieces. These would be defined primarily by a control polygon of vertices $\underline{\mathbf{d}}_i$. This control polygon determines the rough shape of the curve. The exact shape will be fixed after we determine the interior Bézier points $\underline{\mathbf{b}}_{3i\pm 1}$. A convenient way to do this is by setting

$$\underline{\mathbf{b}}_{3i-1} = (1-\alpha_i)\underline{\mathbf{d}}_{i-1} + \alpha_i \underline{\mathbf{d}}_i \tag{9.4}$$

and

$$\underline{\mathbf{b}}_{3i+1} = (1-\beta_i)\underline{\mathbf{d}}_i + \beta_i \underline{\mathbf{d}}_{i+1}. \tag{9.5}$$

Once the α_i and β_i are given, one may choose $\underline{\mathbf{q}}_0$ and $\underline{\mathbf{q}}_1$ and then employ the construction from the previous section, for all curve segments.

We may also find a knot sequence with respect to which our curve is C^2. This is done exactly as in the preceding section. We may then reparametrize the curve using a Moebius transformation; see Figure 9.4.

The α_i and β_i may be viewed as *shape parameters*: they can be used to "fine-tune" the curve shape after the control polygon is specified.

A personal remark: shape parameters promise more flexibility for a designer. In reality, they offer confusion — it is much too tedious to adjust a large number of parameters. This is not to say that methods involving shape parameters are bad, it just points out a deficiency: a method that involves shape parameters is not complete as long as no guidelines are provided that automatically assign reasonable shape parameter values. In that sense, the above method is incomplete.

It is not hard to make this projective spline curve into a rational one in affine space: just use the standard recipe of projecting from projective into affine space. We may apply the concepts of C^2 continuity and reparametrization to these splines, just as we could in Section 9.2.

Another approach for achieving C^2 continuity for G^2 splines is due to W. Degen.[3]

The curve scheme that we described here is equivalent to one devised by W. Boehm, although in a more algebraic form; see [22]. A different approach, based on the concept of so-called β-splines, was taken by Hohmeyer and Barsky; see [67].

9.4 Problems

1. Verify that the NURB curve from Section 15.2 can be formulated as a projective spline.
2. Then formulate a whole circle as a closed projective spline with four segments.
3. Experiment with different choices for α_i and β_i in (9.4) and (9.5) in order to achieve an "optimal shape" of your spline curve.

[3]CAGD 5(3):259–268, 1988

10

Rational B-splines

We generalized Bézier curves to the rational case – using the same approach, we obtain rational B-splines. Instead of "rational B-splines," one typically encounters the term "NURBS." The term has gained too much popularity to be changed, but it is not a good term. In the literature on B-splines (e.g., de Boor [35], [36] or Schumaker [102]), B-splines are introduced as certain functions over an arbitrary partition of the real line. The case of a *uniform* partition is merely a simple subcase, and its negation (NonUniform) should not have been used in a definition.[1]

We have already considered rational quadratic and cubic B-splines in Chapters 6 and 9, respectively. Now we will generalize to arbitrary degrees – a generalization of integral B-splines that will not be very hard at all.

10.1 Definitions

A rational Bézier curve is defined as

$$\mathbf{b}(t) = \frac{\sum_{i=0}^{n} w_i \mathbf{b}_i B_i^n(t)}{\sum_{i=0}^{n} w_i B_i^n(t)}; \quad \mathbf{b}_i \in \mathbb{E}^3,$$

[1]Knots are real numbers, not integers, in general. Imagine someone coined the term "NUNIBS" for Non-Uniform Non-Integer B-splines!

meaning that it is the projection of an integral Bézier curve in $\mathbb{P}^3$. We define rational B-spline curves $\mathbf{s}(u)$ of degree n in exactly the same way:

$$\mathbf{s}(u) = \frac{\sum_{i=0}^{L+n-1} w_i \mathbf{d}_i N_i^n(u)}{\sum_{i=0}^{L+n-1} w_i N_i^n(u)}, \quad \mathbf{d}_i \in \mathbb{E}^3, \tag{10.1}$$

that is to say, it is the projection of the integral B-spline curve $\underline{\mathbf{s}}(u)$:

$$\underline{\mathbf{s}}(u) = \sum_{i=0}^{L+n-1} \underline{\mathbf{d}}_i N_i^n(u); \quad \underline{\mathbf{d}}_i \in \mathbb{P}^3. \tag{10.2}$$

We follow the B-spline notation of [47]: our B-spline curve is defined over a *knot sequence* $\{u_0, \ldots, u_{L+2n-2}\}$, but is only evaluated over the $[u_{n-1}, \ldots, u_{L+n-1}]$. If all knots u_i are simple, i.e., they are listed only once in the knot sequence, then L is the number of curve segments. For each multiple listing of one of the u_i, the number of curve segments decreases by one, and so does the differentiability of the curve.

Typically, the end knots are of multiplicity n: $u_0 = \ldots = u_{n-1}$ and $u_{L+n-1} = \ldots = u_{L+2n-2}$. In this case, the first and last control points, $\underline{\mathbf{d}}_0$ and $\underline{\mathbf{d}}_{L+n-1}$, are on the curve. This is the most desirable configuration for the end knots, as it allows more direct control over the curve shape. In the IGES data specification (Section 15.1), the first and last knot would have to be specified $n+1$ times each. If the end knots are not of multiplicity n, the first and last control points are not on the curve and are called "phantom vertices" [15].

The effect of the weights w_i is demonstrated in Figure 10.1. Each point on the curve is associated with a parameter value u. We can also associate each control point $\underline{\mathbf{d}}_i$ with a parameter value ξ_i, defined as

$$\xi_i = \frac{1}{n}(u_i + \ldots + u_{i+n-1}).$$

The ξ_i are called *Greville abscissae*, see [47].

The basis functions $N_i^n(u)$ are defined as follows: each N_i^n is nonzero over the interval $[u_{i-1}, u_{i+n}]$. This defines the N_i^n up to a scaling factor, which is usually fixed such that all N_i^n form a partition of unity, that is to say

$$\sum_{i=0}^{L+n-1} w_i N_i^n(u) \equiv 1.$$

It follows that for the case of all w_i being unity, the rational B-spline curve is actually an integral one.

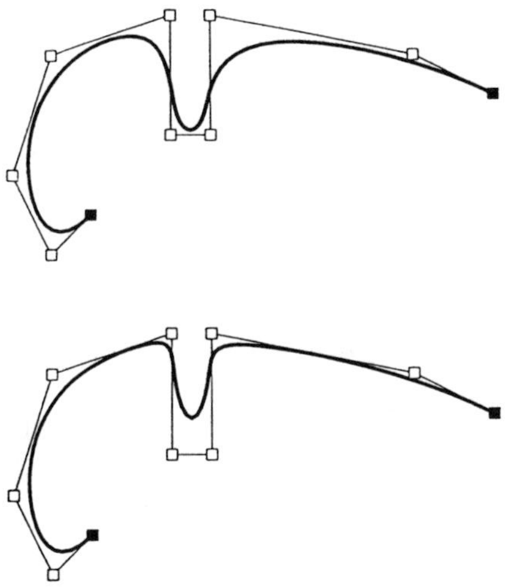

Figure 10.1. NURB weights: top, an integral NURB curve; bottom, after two weights were set to 0.1.

The basis functions satisfy the following recursion:

$$N_l^n(u) = \frac{u - u_{l-1}}{u_{l+n-1} - u_{l-1}} N_l^{n-1}(u) + \frac{u_{l+n} - u}{u_{l+n} - u_l} N_{l+1}^{n-1}(u). \tag{10.3}$$

We may rewrite (10.1) as

$$\mathbf{s}(u) = \sum_{i=0}^{L+n-1} \mathbf{d}_i \frac{w_i N_i^n(u)}{\sum_{j=0}^{L+n-1} w_j N_j^n(u)}. \tag{10.4}$$

In this form, each control point $\mathbf{d}_i$ is multiplied by a rational function $R_j^n(u)$:

$$R_j^n(u) = \frac{w_i N_i^n(u)}{\sum_{j=0}^{L+n-1} w_j N_j^n(u)}. \tag{10.5}$$

We may again introduce *weight points* $\mathbf{q}_i$ by setting

$$\mathbf{q}_i = \frac{w_i \mathbf{d}_i + w_{i+1} \mathbf{d}_{i+1}}{w_i + w_{i+1}}.$$

Each weight point is associated with the parameter value $\zeta_i = \frac{1}{2}\xi_i + \frac{1}{2}\xi_{i+1}$.

These weight points may be used for shape control in exactly the same way as in the case of rational Bézier curves. If some of the weights are negative, the weight points will not be between their corresponding control points. In these cases, singularities are possible.

We may also *reparametrize* NURBS in the same sense as we could rational Bézier curves or the projective splines from Chapter 9. Let a Moebius transformation map the knot sequence $\{u_i\}$ to a new knot sequence $\{\hat{u}_i\}$, such that the first and last knots are not changed. The weight point parameter values ζ_i will be mapped to new values $\hat{\zeta}_i$. The weight points themselves will be mapped to new weight points $\hat{\mathbf{q}}_i$, and the following identity must hold:

$$\mathrm{cr}(\mathbf{d}_i, \mathbf{q}_i, \hat{\mathbf{q}}_i, \mathbf{d}_{i+1}) = \mathrm{cr}(\xi_i, \zeta_i, \hat{\zeta}_i, \xi_{i+1}). \tag{10.6}$$

This is in analogy to the Bézier case (7.22), but now things are more complicated: in (7.22), the right-hand side was the same for all i, whereas now it does depend on i! NURB reparametrization was first addressed by E. Lee and M. Lucian [77].[2]

An illustration is given by Figure 10.2 although developed in a different context, projective splines are C^2 piecewise projective cubics once a suitable knot sequence has been chosen. Projected into affine space, they then are C^2 cubic NURBS.

10.2 Knot Insertion and the de Boor Algorithm

The basic operation for a B-spline curve is the process of *knot insertion*, devised by W. Boehm in 1981 [20]. This is best described in terms of homogeneous coordinates: suppose we want to insert a new knot u into our knot sequence, thus splitting an existing curve segment into two new ones. Once we picked u, there exists an interval number I such that $u_I \le u < u_{I+1}$. The new control points $\underline{\mathbf{d}}_i^u$ are given by

$$\underline{\mathbf{d}}_i^u = \frac{u_{i+n-1} - u}{u_{i+n-1} - u_{i-1}} \underline{\mathbf{d}}_{i-1} + \frac{u - u_{i-1}}{u_{i+n-1} - u_{i-1}} \underline{\mathbf{d}}_i; \quad i = I - n + 2, \ldots, I + 1. \tag{10.7}$$

If more than one knot is to be inserted into the curve's domain, the knot insertion procedure is applied repeatedly. It is important to note that it does not matter in which order these knots are inserted ([47]). For the simultaneous insertion of several knots, a rivaling technique is the *Oslo algorithm*; see [32], [82], [95].

Knot insertion can be used to *evaluate* B-spline curves; that process is

[2] That article corrected a false statement made by myself in [46].

called the *de Boor algorithm* and proceeds as follows (in $\mathbb{P}^3$):

$$\underline{\mathbf{d}}_i^k(u) = (1-\alpha_i^k)\underline{\mathbf{d}}_{i-1}^{k-1}(u) + \alpha_i^k \underline{\mathbf{d}}_i^{k-1}(u), \left\{ \begin{array}{l} k = 1, \ldots, n-r, \\ i = I-n+k+1, \ldots, I+1, \end{array} \right. \tag{10.8}$$

with

$$\alpha_i^k = \frac{u - u_{i-1}}{u_{i+n-k} - u_{i-1}}.$$

Then $\underline{\mathbf{d}}_{I+1}^{n-r}(u)$ is the value of the B-spline curve at parameter value u. Here, r denotes the multiplicity of u in case it was already one of the knots — if it was not, we set $r = 0$. This algorithm is due to C. de Boor[35].

In the rational case, we have to modify the de Boor algorithm in a way that should by now be obvious:

Given:

- curve degree n,
- knots $u_0, \ldots, u_{L+2n-2}$,
- L denoting the number of all curve intervals, possibly including zero length ones,
- control points $\mathbf{d}_0, \ldots, \mathbf{d}_{L+n-1}$,
- weights $w_0, \ldots, w_{L+n-1}$
- $u \in [u_I, u_{I+1}]; n-1 \le I < L+n-1$, the evaluation parameter,
- multiplicity r should u be a knot having that multiplicity. If u is not one of the knots, set $r = 0$.

Find: $\mathbf{s}(u)$ as given by (10.1).

Do: Set $\underline{\mathbf{d}}_i = w_i \mathbf{d}_i; i = 0, \ldots, L+n-1$, apply (10.8) to compute $\underline{\mathbf{d}}_{I+1}^n(u)$ and then set

$$\mathbf{s}(u) = \frac{\mathbf{d}_{I+1}^n(u)}{w_{I+1}^n(u)}. \tag{10.9}$$

We note that one could also project any of the intermediate $\underline{\mathbf{d}}_i^k$ into affine space; this is done in the next section.

For an illustration of the quadratic case, see Figure 10.2.

10.3 Derivatives

Integral B-spline curves of degree n are $n-r$ times differentiable at a knot that has multiplicity r. At locations other than the knots, they are infinitely often differentiable, since there they behave like integral polynomials. The

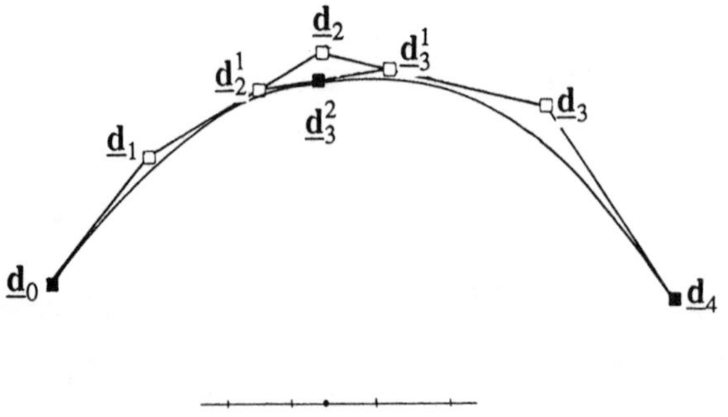

Figure 10.2. The de Boor algorithm: an example for the quadratic case $n = 2$

rational case is more complicated, of course: derivatives may not exist at singularities, and they become very involved for higher derivatives.

While for higher derivatives the recursive formula (7.31) works,[3] we are luckier for the first derivative.

The de Boor algorithm for the homogeneous points $\underline{\mathbf{d}}_i$ yields the curves' tangent: it is spanned by $\underline{\mathbf{d}}_I^{n-1} \wedge \underline{\mathbf{d}}_{I+1}^{n-1}$. The derivative of the projective curve is given by

$$\dot{\underline{\mathbf{s}}}(u) = \frac{n}{u_{i+1} - u_I}\left[\underline{\mathbf{d}}_{I+1}^{n-1} - \underline{\mathbf{d}}_I^{n-1}\right].$$

Unfortunately, it does not project to the derivative of the rational curve. But following a result by Floater [53], we have

$$\dot{\underline{\mathbf{s}}}(u) = \frac{n}{u_{I+1} - u_I}\frac{w_I^{n-1} w_{I+1}^{n-1}}{[w_{I+1}^n]^2}\left[\mathbf{d}_{I+1}^{n-1} - \mathbf{d}_I^{n-1}\right]. \qquad (10.10)$$

Each intermediate point $\mathbf{d}_i^k$ depends on u, and thus traces out a curve. Its derivative $\dot{\mathbf{d}}_i^k(u)$ is given by

$$\dot{\mathbf{d}}_i^k(u) = \frac{k}{u_{i-n+k} - u_{i-1}}\frac{w_{i-1}^{k-1} w_i^{k-1}}{[w_i^k]^2}\left[\mathbf{d}_i^{k-1} - \mathbf{d}_{i-1}^{k-1}\right],$$

thus generalizing (10.10), a result also due to Floater.

This may be used to obtain a *bound* on the derivatives of a NURB curve, quite in analogy to (7.34):

$$||\dot{\mathbf{s}}(u)|| \le n\left[\frac{\max\{w_i\}}{\min\{w_i\}}\right]^2 \max_{0<i\le L+n-1}\left\{\frac{||\mathbf{d}_i - \mathbf{d}_{i-1}||}{u_{i+n-1} - u_{i-1}}\right\}. \qquad (10.11)$$

[3] We just have to interpret the involved $\mathbf{p}(t), w(t), \mathbf{b}(t)$ as B-splines.

10.4 Properties

Rational B-splines have properties that should not be surprising at this point; we list the most important ones:

Projective invariance:

The following two operations yield the same result:

a) Evaluate a NURB curve for a parameter value u, yielding a point $\mathbf{s}(u)$, then transform $\mathbf{s}(u)$ by a projective map to a point $\Phi\mathbf{s}(u)$.

b) Transform the control polygon plus weights by Φ, then evaluate at u.

Note that Step b) will change the weights. In case the projective transformation was actually affine, we retain affine invariance: the weights will not change.

Convex hull property:

If the weights are nonnegative, the curve lies in the convex hull of the control polygon. In fact, any point on the curve lies in the convex of just $n+1$ consecutive control points.

Variation diminishing property:

This holds provided the weights are nonnegative.

Superset property:

NURBS are a true superset of rational Bézier curves. In order to write a rational Bézier curve of degree n as a NURB curve, define a knot sequence

$$\{u_0, u_1, \ldots, u_{2n-1}\} = \{0^{<n>}, 1^{<n>}\},$$

where $\cdot^{<n>}$ denotes n–fold repetition. Note that the Greville abscissae now satisfy $\xi_i = i/n$.

10.5 Conversion to Rational Bézier Form

Ultimately, every NURB curve is an ordered collection of rational polynomial curve segments. Each of these may be represented in rational Bézier form. The corresponding Bézier points and weights are again found by the process of knot insertion: just insert each knot until its multiplicity is n, where n is the degree of the curve. Thus the NURB/rational Bézier conversion is nothing but a repeated application of (10.7). Figure 10.3 gives an example.

If we are only interested in some particular segment, $[u_I, u_{I+1}]$ say, then the conversion may be carried out strictly locally. Simply insert u_I until it is of full multiplicity and then insert u_{I+1} until it is of full multiplicity.

Some remarks concerning the relationship between rational Bézier curves and NURBS:

- Suppose a particular curve segment has to be evaluated at many parameter values. Applying the de Boor algorithm many times would be slower than applying the de Casteljau algorithm many times, even counting in the overhead of the initial conversion.
- The piecewise Bézier polygon hugs the curve closer than its B-spline polygon. Thus for interference checks (as needed for clipping or intersection algorithms), the piecewise Bézier form is preferable.
- A rational Bézier curve is a special case of NURBS: if the knots are

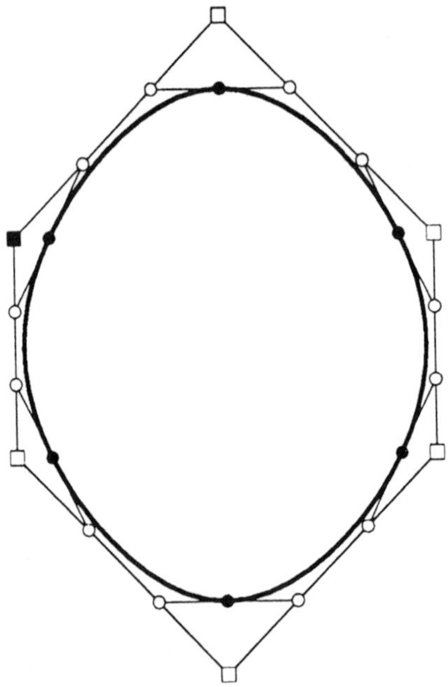

Figure 10.3. NURB conversions: a cubic NURB curve is converted to rational cubic Bézier form.

$\{a^{<n>}, b^{<n>}\}$, then the NURB curve is just a rational Bézier curve of degree n over $[a, b]$.

10.6 Problems

1. Find a recursion, similar to (10.3) from (10.5) for the functions R_j^n.

2. Find a formula similar to (7.33) for NURBS. How does it compare to (10.11)?

3. Is a NURBS curve in the convex hull of $\mathbf{d}_0, \mathbf{q}_0, \mathbf{q}_1, \dots, \mathbf{q}_{L+n-2}, \mathbf{d}_{L+n-1}$, in analogy to the Bézier case as illustrated in Figure 7.2?

11

Rectangular Patches

The theory of rational curves was an easy and straightforward extension of our treatment of conics. In the surface case, things will turn out to be more complicated. It would appear natural to base the development of surfaces on an extension of quadrics, but this is not possible. We therefore start with the parametric definition of a rational surface, shifting between affine and projective contexts as appropriate. We introduce rational Bézier patches and NURBS just as we did for the curve case. Again, the Bézier form is the basic building block, and we start from it.

11.1 Bilinear Patches

A point $\mathbf{x}(u, v)$ on a rational bilinear patch is the image of a point (u, v), defined by

$$\mathbf{x}(u, v) = \frac{\sum_{i=0}^{1} \sum_{j=0}^{1} w_{ij} \mathbf{b}_{ij} B_i^1(u) B_j^1(v)}{\sum_{i=0}^{1} \sum_{j=0}^{1} w_{ij} B_i^1(u) B_j^1(v)}. \tag{11.1}$$

It may be viewed as a projection of a bilinear polynomial patch in $\mathbb{P}^3$, given by

$$\underline{\mathbf{x}}(u, v) = \sum_{i=0}^{1} \sum_{j=0}^{1} \underline{\mathbf{b}}_{ij} B_i^1(u) B_j^1(v). \tag{11.2}$$

This patch is a map of a quadrilateral in the projective plane, defining a Moebius net. The term *bilinear* is due to the fact that the patch equation is linear in u as well as in v. Geometrically, this means that the patch contains two families of straight lines: at a fixed point $\mathbf{x}(u^*, v^*)$, one line is given by

$$\underline{\mathbf{l}}(u) = (1-u)\big[(1-v^*)\underline{\mathbf{b}}_{00} + v^*\underline{\mathbf{b}}_{01}\big] + u\big[(1-v^*)\underline{\mathbf{b}}_{10} + v^*\underline{\mathbf{b}}_{11}\big],$$

the other one by

$$\underline{\mathbf{l}}(v) = (1-v)\big[(1-u^*)\underline{\mathbf{b}}_{00} + u^*\underline{\mathbf{b}}_{10}\big] + v\big[(1-u^*)\underline{\mathbf{b}}_{01} + u^*\underline{\mathbf{b}}_{11}\big].$$

These two lines span a plane, which is the *tangent plane* of our patch.

Figure 11.1 illustrates. Since intersecting straight lines are preserved under projections, the rational bilinear patch (11.1) also contains two families of straight lines. Such surfaces are called *doubly ruled.*

Let us now investigate some of the geometric properties of our patch, resorting to $\mathbb{P}^3$ for convenience. Let (u, v) trace out a straight line in the domain, i.e., we set $u = u(t)$ and $v = v(t)$, and both $u(t)$ and $v(t)$ are linear in t. Inserting this into (11.2), we obtain a curve $\underline{\mathbf{c}}(t) = \underline{\mathbf{x}}(u(t), v(t))$, lying completely on the patch. It does not take much effort to realize that $\mathbf{c}(t)$ is quadratic in t. Note that the two straight lines through a point are a special case of this result.

We thus have: straight lines in the domain of a bilinear patch are mapped to conics. Since these conics are planar curves, they may be viewed as the intersection curves between the patch and a set of planes. Now a natural question is: are *all* planar sections of a bilinear patch conic sections? They

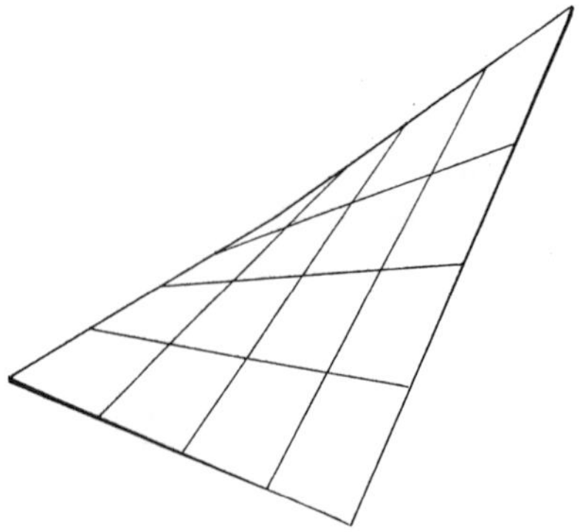

Figure 11.1. Bilinear patches: a patch is covered by two families of straight lines.

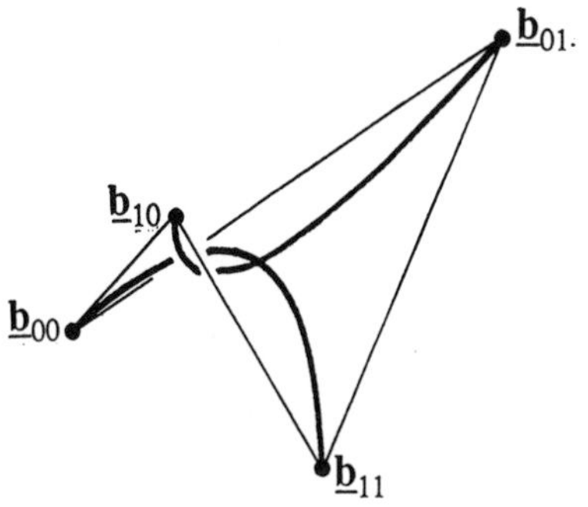

Figure 11.2. Bilinear patches: the two diagonal curves.

are indeed, although we give no proof here. Note that *every* plane intersects the bilinear patch, since it intersects every line in the two families of straight lines!

Thus a rational bilinear patch is a *quadric*, since a quadric is defined by the fact that any of its planar sections yields a conic. According to Section 13.1, it is either a hyperbolic paraboloid — then it is tangent to the plane at infinity; or it is a hyperboloid of one sheet — then it intersects the plane at infinity. The special case of all $w_{ij} = 1$, i.e., the integral case, yields the hyperbolic paraboloid.

A look at the "diagonals" clarifies this observation. Let $\underline{\mathbf{d}}_1(t) = \underline{\mathbf{x}}(t,t)$ and $\underline{\mathbf{d}}_2(t) = \underline{\mathbf{x}}(t, 1-t)$ be the two diagonal curves. Each of these curves is a conic and has the following Bézier points:

$$\begin{aligned} \underline{\mathbf{d}}_1 : &\quad \underline{\mathbf{b}}_{00}, (\underline{\mathbf{b}}_{01} + \underline{\mathbf{b}}_{10})/2, \underline{\mathbf{b}}_{11} \\ \underline{\mathbf{d}}_2 : &\quad \underline{\mathbf{b}}_{01}, (\underline{\mathbf{b}}_{00} + \underline{\mathbf{b}}_{11})/2, \underline{\mathbf{b}}_{10}. \end{aligned}$$

It is easy to verify that the two diagonal curves pass through the point $\underline{\mathbf{x}}(\frac{1}{2}, \frac{1}{2})$; see Figure 11.2 for an illustration.

But they meet one more time: we also have that

$$\underline{\mathbf{d}}_1(\infty) \hat{=} \underline{\mathbf{d}}_2(\infty) \hat{=} \underline{\mathbf{b}}_{00} - \underline{\mathbf{b}}_{01} - \underline{\mathbf{b}}_{10} + \underline{\mathbf{b}}_{11},$$

which is verified by a simple exercise in algebra.[1]

After projecting into affine space, when we have a rational patch (11.1) with weights w_{ij}, the diagonal curves will be of one of the three conic types.

[1]It is interesting to note that this equals the mixed partial derivative $\underline{\mathbf{x}}_{uv}$, or the twist of the bilinear patch!

They can either be both parabolas or they can be one hyperbola and one ellipse. In the first case, the patch *touches* the plane at infinity, and we have the case of a hyperbolic paraboloid. In the second case, the plane at infinity is *intersected*, namely by the diagonal hyperbola, and we have a hyperboloid of one sheet.

11.2 Bézier Patches

A rational rectangular Bézier patch is given by

$$\mathbf{x}(u,v) = \frac{\sum_{i=0}^{m}\sum_{j=0}^{n} w_{ij}\mathbf{b}_{ij}B_i^m(u)B_j^n(v)}{\sum_{i=0}^{m}\sum_{j=0}^{n} w_{ij}B_i^m(u)B_j^n(v)}; \qquad 0 \le u,v \le 1. \tag{11.3}$$

It is the image of an integral projective patch:

$$\underline{\mathbf{x}}(u,v) = \sum_{i=0}^{m}\sum_{j=0}^{n} \underline{\mathbf{b}}_{ij}B_i^m(u)B_j^n(v). \tag{11.4}$$

The points $\underline{\mathbf{b}}_{ij}$ or $\mathbf{b}_{ij}$ are called the *control points* of the surface; they form the *control net*. The w_{ij} of the rational patch are called *weights*. The control points are arranged in a rectangular pattern; here is the bicubic case:

$$\begin{matrix} \underline{\mathbf{b}}_{00}\underline{\mathbf{b}}_{10}\underline{\mathbf{b}}_{20}\underline{\mathbf{b}}_{30} \\ \underline{\mathbf{b}}_{01}\underline{\mathbf{b}}_{11}\underline{\mathbf{b}}_{21}\underline{\mathbf{b}}_{31} \\ \underline{\mathbf{b}}_{02}\underline{\mathbf{b}}_{12}\underline{\mathbf{b}}_{22}\underline{\mathbf{b}}_{32} \\ \underline{\mathbf{b}}_{03}\underline{\mathbf{b}}_{13}\underline{\mathbf{b}}_{23}\underline{\mathbf{b}}_{33} \end{matrix}$$

Bézier patches are well understood, and there is a vast amount of literature on them, see [47], [70]. Instead of rederiving every known result for the rational case, we just state some of the fundamental properties:

- If all weights are positive, the rational patch lies in the convex hull of its control net. If we allow the parameters to take on values outside $[0, 1]$, the corresponding surface points will no longer be in this convex hull.

- The four boundary curves of the patch are defined by the control net boundary polygons and their weights.

- In particular, the four control net corners lie on the patch.

- Every isoparametric curve $u = const$ or $v = const$ is a rational curve of degree m or n, respectively.

- The "diagonal" $\mathbf{x}(t, t)$ is a rational curve of degree $m + n$.

- A straight line may have up to $2mn$ intersections with the patch. This number is referred to as its *algebraic degree.*
- A rational patch is *degree elevated* by degree elevating all its rows or columns according to Section 7.3.
- A rational patch is *subdivided* by interpreting all of its control polygon rows (or columns) as rational curve control polygons and then subdividing them, following the subdivision process outlined in Section 7.9.

Figure 11.3 shows two rational bicubics, both defined by the same control net but with different weights. As in the curve case, increasing a weight pulls the surface closer to the corresponding control point, decreasing it results in a "push back" effect. In the limiting case of one weight — but none of its neighbors — tending to infinity, the rational surface will pass through the corresponding control point.[2]

In the figure, we use *isophotes* to highlight patch shape changes. Isophotes are areas on a surface where the normal vector does not differ by more than a prescribed tolerance from a fixed vector. Isophotes[3] are a very effective tool for detecting shape features in a surface — much more so than shaded images. They take some time to get used to, but once mastered, they are an indispensable tool for surface design.

11.3 De Casteljau Algorithms, Blossoms, and Derivatives

The de Casteljau algorithm for Bézier curves carries over to surfaces in two different ways. For a description, it is more convenient to stay in projective space.

First, let us rewrite (11.4) as

$$\underline{\mathbf{x}}(u,v) = \sum_{i=0}^{m} \Big[\sum_{j=0}^{n} \underline{\mathbf{b}}_{ij} B_j^n(v)\Big] B_i^m(u).$$

Denoting the terms in square brackets by $\underline{\mathbf{c}}_i(v)$, we can now formulate our surface evaluation algorithm as:

1. For each i, only consider the ith column of control points and treat them as a curve polygon. Perform a curve de Casteljau algorithm with respect to v, yielding a point $\underline{\mathbf{c}}_i(v)$.

[2]It does not matter if the weight tends to $+\infty$ or to $-\infty$, the surface will approach the control point from different directions, but still pass through it in the limit.

[3]Isophotes are a special case of *reflection lines*, which simulate the reflection of a straight line in a surface.

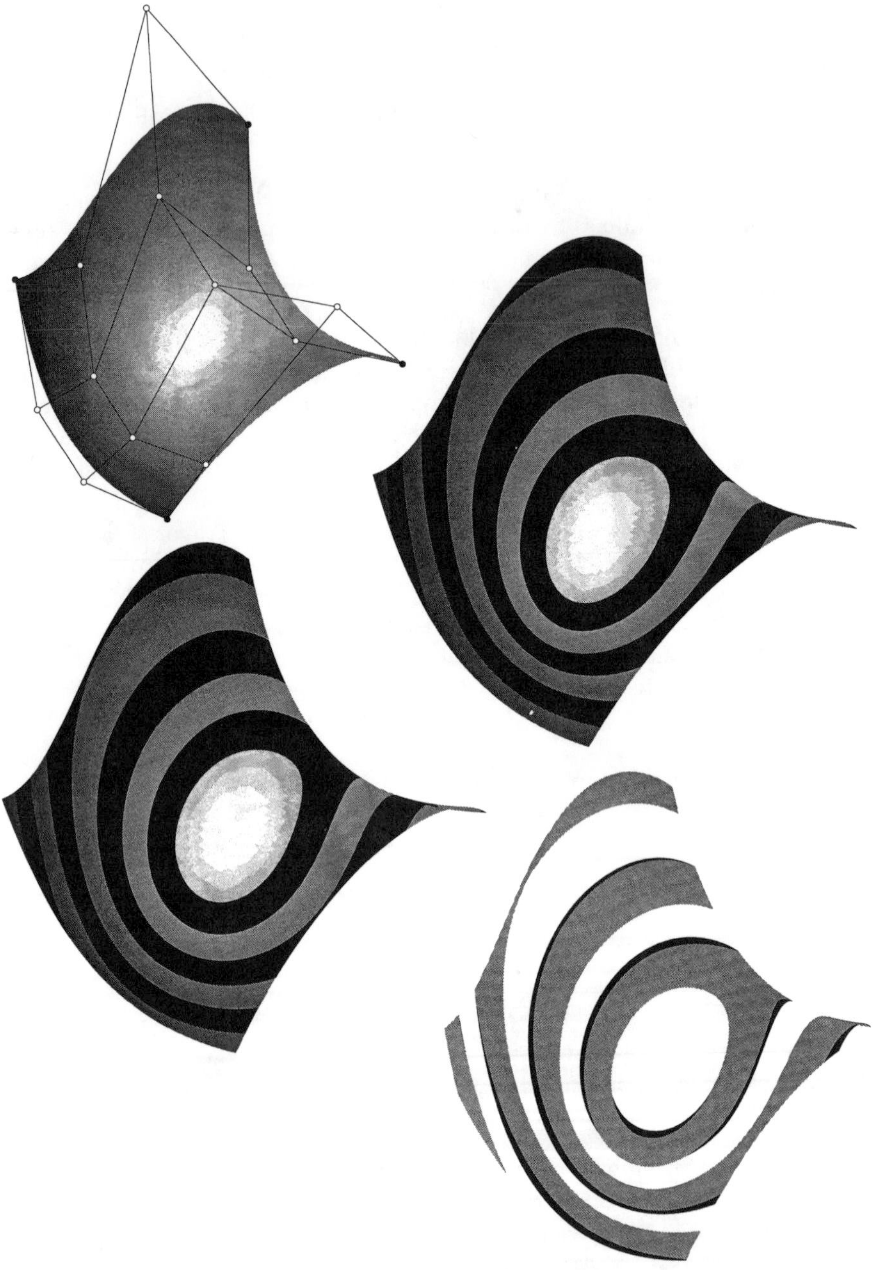

Figure 11.3. Rational bicubics: top left: an integral patch. Top right: a set of isophotes. Bottom left: isophotes after $w_1 = 1$ was changed to 2.0. Bottom right: the difference plot of the two sets of isophotes.

2. Interpret the resulting points $\underline{\mathbf{c}}_i(v)$ as a control polygon for a Bézier curve, and perform one more de Casteljau algorithm with respect to u. The result is the desired surface point.

Figure 11.4 illustrates this approach.

The points $\underline{\mathbf{c}}_i(v)$ are the control points of the isoparametric curve corresponding to the fixed parameter value u. The final de Casteljau algorithm in the u–direction yields the tangent to the isoparametric curve, and thus a tangent to the surface.

We could have organized our approach differently: we might have started by evaluating in the u-direction first, and then in the v-direction. That way, we could have found a tangent in the v-direction of the surface. Many applications of surface evaluations do not only need the point on the surface, but also its tangent plane. We then need tangents in both the u- and the v-direction. Instead of running our algorithm twice, we reorganize our evaluation strategy by employing *blossoms.*

The blossom

$$\underline{\mathbf{b}}[u_1, \ldots, u_m \mid v_1, \ldots, v_n]$$

of a rectangular patch is defined by first computing the blossom value $[u_1, \ldots, u_m]$ for every row, and then using these values as Bézier control points for a blossom value $[v_1, \ldots, v_n]$ in the v-direction. Of course, we could have computed it the other way around: first v, then u, as well!

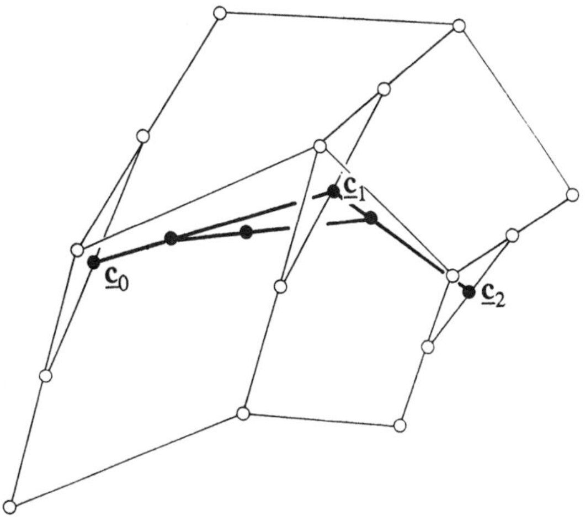

Figure 11.4. Surface evaluation: in this example, three de Casteljau algorithms are performed in the v- direction, then one in the u-direction.

In case all u_i equal a value u and all v_j equal a value v, then the blossom yields a point on the surface:

$$\underline{\mathbf{x}}(u,v) = \underline{\mathbf{b}}[u^{<m>} \mid v^{<n>}].$$

The Bézier points may also be obtained as special blossom values:

$$\underline{\mathbf{b}}_{i,j} = \underline{\mathbf{b}}[0^{<i>}, 1^{<m-i>} \mid 0^{<j>}, 1^{<n-j>}], \tag{11.5}$$

almost exactly as in the curve case!

An *isoparametric curve* in, say, the u-direction is defined by keeping v fixed at a value $\bar{v}$, and by letting u vary. In projective space, the control points $\underline{\bar{\mathbf{b}}}_i$ of this curve are conveniently computed to

$$\underline{\bar{\mathbf{b}}}_i = [0^{<m-i>}, 1^{<i>} \mid v^{<n>}].$$

This means that each column of control points is to be interpreted as a degree n Bézier control polygon and has to be evaluated at $\bar{v}$, yielding the $\underline{\bar{\mathbf{b}}}_i$.

The corresponding rational curve is obtained in the usual way: with $\bar{w}_i$ being the fourth coordinate of $\underline{\bar{\mathbf{b}}}_i$, the rational curve's control points $\bar{\mathbf{b}}_i$ are given by $\bar{\mathbf{b}}_i = \underline{\bar{\mathbf{b}}}_i / \bar{w}_i$, and their weights are the $\bar{w}_i$.

Needless to say, we can compute the control points of isoparametric curves in the v-direction in a completely analogous way!

The control points of an isoparametric curve are a by-product of the evaluation process. We may use them to compute a partial derivative along the curve by employing (7.32). This only gives one partial; for the other one, we would have to re-evaluate, exchanging the order of u- and v-evaluation. This is tedious, and in case both partials are desired, the following is more efficient.

The *bilinear osculant* at $\underline{\mathbf{x}}(u,v)$ is the bilinear patch given by four Bézier points $\underline{\mathbf{a}}_{i,j}$, which we express in blossom notation:

$$\begin{bmatrix} \underline{\mathbf{a}}_{0,0} & \underline{\mathbf{a}}_{0,1} \\ \underline{\mathbf{a}}_{1,0} & \underline{\mathbf{a}}_{1,1} \end{bmatrix} = \begin{bmatrix} \underline{\mathbf{b}}[u^{<m-1>}, 0 \mid v^{<n-1>}, 0] & \underline{\mathbf{b}}[u^{<m-1>}, 0 \mid v^{<n-1>}, 1] \\ \underline{\mathbf{b}}[u^{<m-1>}, 1 \mid v^{<n-1>}, 0] & \underline{\mathbf{b}}[u^{<m-1>}, 1 \mid v^{<n-1>}, 1] \end{bmatrix}; \tag{11.6}$$

see Figure 11.5. Of course, it does not matter in which order the blossoms are computed.

Figure 11.6 illustrates one strategy to compute the bilinear osculant for the case $m = 3,\ n = 2$.

The bilinear osculant with these four control points agrees with the original surface in u-partial, v-partial, and twist. In particular, it therefore agrees with the surface in its tangent plane.

Higher derivatives in projective space[4] may also be computed using the blossoming technique:

[4]Recall: these are *points*!

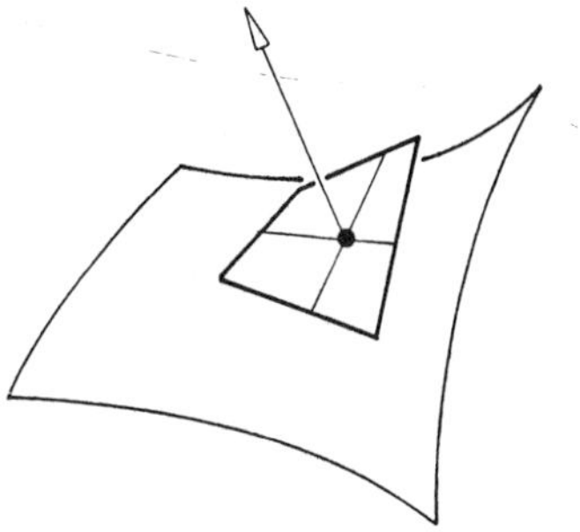

Figure 11.5. The bilinear osculant: it agrees with the given surface in u- and v-partials.

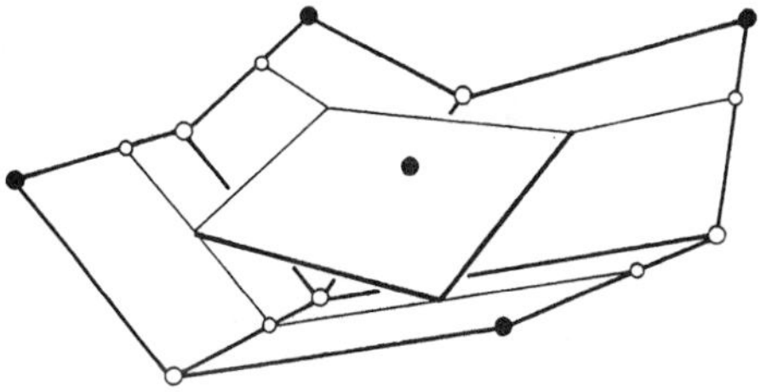

Figure 11.6. The bilinear osculant: several de Casteljau steps for its computation are shown.

$$\frac{\partial^{r+s}}{\partial u^r \partial v^s}\underline{\mathbf{x}}(u,v) \hat{=} \underline{\mathbf{b}}[u^{<m-r>}, \infty^{<r>} \mid v^{<n-s>}, \infty^{<s>}]. \tag{11.7}$$

Compare with (7.38)!

A point on a rational patch may be computed by performing the above projective evaluation algorithms and then projecting the result into affine space, or by projecting every step of those algorithms into affine space. In particular, we may compute the rational bilinear osculant to the patch: it is given by the four points

$$\mathbf{b}_{0,0}^{m-1,n-1}, \mathbf{b}_{0,1}^{m-1,n-1}, \mathbf{b}_{1,0}^{m-1,n-1}, \mathbf{b}_{1,1}^{m-1,n-1}$$

and their respective weights

$$w_{0,0}^{m-1,n-1}, w_{0,1}^{m-1,n-1}, w_{1,0}^{m-1,n-1}, w_{1,1}^{m-1,n-1}.$$

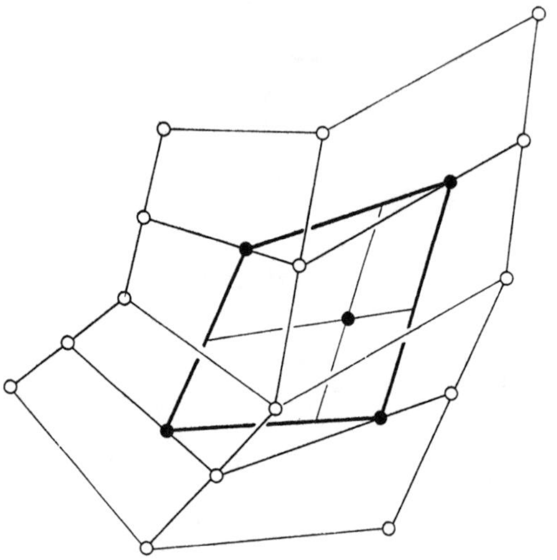

Figure 11.7. Partials: the involved control points on the bilinear osculant.

We may now use this rational bilinear osculant to obtain partials:

$$\mathbf{x}_u(u,v) = m\frac{w_{0,0}^{m-1,n}w_{1,0}^{m-1,n}}{[w_{0,0}^{m,n}]^2}\left[\mathbf{b}_{1,0}^{m-1,n} - \mathbf{b}_{0,0}^{m-1,n}\right] \tag{11.8}$$

and

$$\mathbf{x}_v(u,v) = n\frac{w_{0,0}^{m,n-1}w_{0,1}^{m,n-1}}{[w_{0,0}^{m,n}]^2}\left[\mathbf{b}_{0,1}^{m,n-1} - \mathbf{b}_{0,0}^{m,n-1}\right]. \tag{11.9}$$

Figure 11.7 explains the notation.

If we want the tangent plane (or, equivalently, the normal) of the corresponding rational surface in affine space, we simply take the cross product of $\mathbf{x}_u$ and $\mathbf{x}_v$.

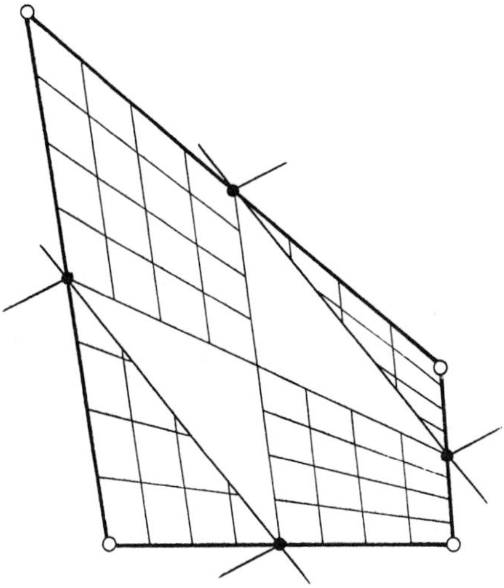

Figure 11.8. Weight points: neighboring control points (open) give rise to coplanar weight points (solid).

11.4 The Weight Points

We may form the averages $\mathbf{p}_{i,j}$ in the i-direction of any two adjacent control points:

$$\mathbf{p}_{i,j} = \frac{w_{i,j}\mathbf{b}_{i,j} + w_{i+1,j}\mathbf{b}_{i+1,j}}{w_{i,j} + w_{i+1,j}}$$

and similarly the averages $\mathbf{q}_{i,j}$ in the j-direction:

$$\mathbf{q}_{i,j} = \frac{w_{i,j}\mathbf{b}_{i,j} + w_{i,j+1}\mathbf{b}_{i,j+1}}{w_{i,j} + w_{i,j+1}}.$$

These *weight points* are shown in Figure 11.8, there for the special case $m = n = 1$ and all $w_{i,j} = 1$. This characterizes a hyperbolic paraboloid — note that for this case, the weight points actually form a parallelogram.

Any subquadrilateral has four weight points on its edges. Although the control weights and control points are arbitrary, the weight points always satisfy one condition: they are *coplanar*.[5] This may be surprising in an affine context, but it is close to trivial in a projective one: there we have

$$\underline{\mathbf{p}}_{i,j} = \underline{\mathbf{b}}_{i,j} + \underline{\mathbf{b}}_{i+1,j}, \quad \underline{\mathbf{q}}_{i,j} = \underline{\mathbf{b}}_{i,j} + \mathbf{b}_{i,j+1}.$$

[5]This was first pointed out to me by H. Pottmann.

Assigning coordinates[6]

$$\underline{\mathbf{b}}_{i,j} = \begin{bmatrix} 1 \\ 0 \\ 0 \\ 0 \end{bmatrix}, \underline{\mathbf{b}}_{i+1,j} = \begin{bmatrix} 0 \\ 1 \\ 0 \\ 0 \end{bmatrix}, \underline{\mathbf{b}}_{i,j+1} = \begin{bmatrix} 0 \\ 0 \\ 1 \\ 0 \end{bmatrix}, \underline{\mathbf{b}}_{i+1,j+1} = \begin{bmatrix} 0 \\ 0 \\ 0 \\ 1 \end{bmatrix},$$

all we need to do is check that the determinant $|\underline{\mathbf{p}}_{i,j}, \underline{\mathbf{p}}_{i,j+1}, \underline{\mathbf{q}}_{i,j}, \underline{\mathbf{q}}_{i+1,j}|$ vanishes. This is a straightforward exercise in algebra.

We have thus lost a useful property of rational Bézier curves: there, we could assign weight points in an intuitive and geometric way — now, the weight points have to satisfy constraints and thus can not be used as independent input.

11.5 Reparametrization

In the curve case, we could change the parametrization by a Moebius transformation without changing the curve shape; the same is true now. The parameters u and v are independent variables; we may subject each of them to a Moebius transformation independently. If the reparametrization constants (see Section 7.5) are c for the parameter u and d for v, then the new patch equation is

$$\underline{\mathbf{x}}(\hat{u}, \hat{v}) = \sum_{i=0}^{m} \sum_{j=0}^{n} c^i d^j \underline{\mathbf{b}}_{i,j} B_i^m(\hat{u}) B_j^n(\hat{v}). \tag{11.10}$$

The corresponding rational case is now trivial:

$$\mathbf{x}(\hat{u}, \hat{v}) = \frac{\sum_{i=0}^{m} \sum_{j=0}^{n} c^i d^j w_{i,j} \mathbf{b}_{i,j} B_i^m(\hat{u}) B_j^n(\hat{v})}{\sum_{i=0}^{m} \sum_{j=0}^{n} c^i d^j w_{i,j} B_i^m(\hat{u}) B_j^n(\hat{v})}. \tag{11.11}$$

Figure 11.9 shows the effect of such a reparametrization on a grid of isoparametric curves of the surface from Figure 11.3.

11.6 Rectangular NURBS Surfaces

Rational Bézier patches were obtained as follows: construct a tensor product surface in projective space, then project into affine space, thus creating

[6] Recall that we may always do this without loss of generality, as long as the four points are not coplanar – in which case we have nothing to prove.

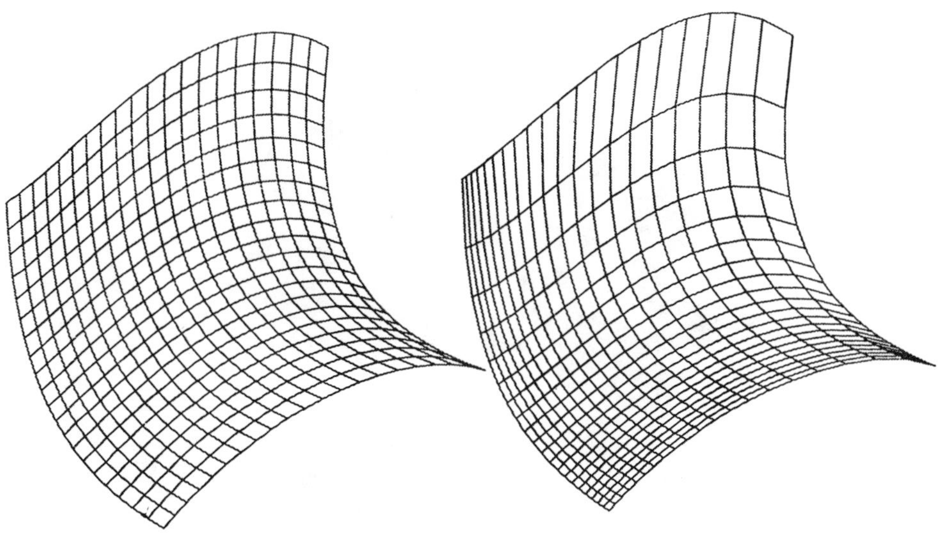

Figure 11.9. Surface reparametrization: left, an integral patch; right, after a reparametrization with $c = 2$, $d = 3$.

a rational surface. The story does not change for B-splines: a projective, piecewise polynomial tensor product B-spline surface is given by

$$\underline{\mathbf{s}}(u,v) = \sum_{i=0}^{L+m-1} \sum_{j=0}^{K+n-1} \underline{\mathbf{d}}_{i,j} N_i^m(u) N_j^n(v), \tag{11.12}$$

where the $N_i^m(u)$ and $N_j^n(v)$ are the univariate B-spline basis functions from Chapter 10. The knot sequence in the u-direction is defined by $\{u_0, \ldots, u_{L+2m-2}\}$; in the v-direction, we have $\{v_0, \ldots, v_{K+2n-2}\}$.

After projection into affine space, we obtain the NURB surface

$$\mathbf{s}(u,v) = \frac{\sum_{i=0}^{L+m-1} \sum_{j=0}^{K+n-1} v_{i,j} \mathbf{d}_{i,j} N_i^m(u) N_j^n(v)}{\sum_{i=0}^{L+m-1} \sum_{j=0}^{K+n-1} v_{i,j} N_i^m(u) N_j^n(v)}; \quad \mathbf{d}_i \in \mathbb{E}^3. \tag{11.13}$$

The weights $v_{i,j}$ are the fourth coordinates of the $\underline{\mathbf{d}}_{i,j}$ – we do not denote them by $w_{i,j}$ to avoid confusion with the Bézier weights. As usual, the $\{\mathbf{d}_{i,j}\}$ are called *control points*, forming the control net. Figure 11.10 shows an example. The weights $v_{i,j}$ may be used to fine tune a given shape. In Figure 11.10, we use isophotes as above to highlight shape features of a surface. After changing some weights, a new pattern emerges. The two

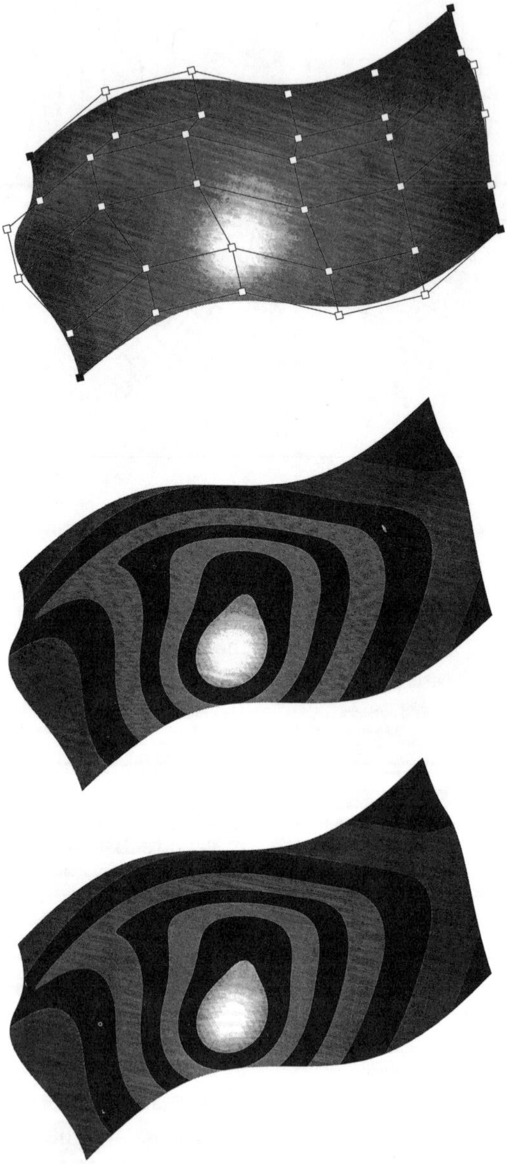

Figure 11.10. NURB surfaces: top, control net and surface; center, isophotes; bottom, isophotes after doubling some weights.

patterns are not markedly different — the only changes are in the right part of the surface. This is a typical situation in surface design: sometimes tiny changes are sufficient to meet a designer's intent.

There are several ways to evaluate NURB surfaces; we discuss the tensor product de Boor algorithm and the conversion to rational Bézier form.

In the case of rectangular Bézier patches, we had several options to apply de Casteljau algorithms for the evaluation of a surface point. The same is true now, except that we replace univariate de Casteljau algorithms by univariate de Boor algorithms. One strategy is as follows: first, we determine in which intervals u and v are, i.e., we find I and J such that $u_I \le u < u_{I+1}$ and $v_J \le v < v_{J+1}$. Then, we take every row of the projective control net $\{\underline{\mathbf{d}}_{i,j}\}$, interpret it as a B-spline curve control polygon, and use the de Boor algorithm (10.8) to evaluate it at u. Then take all points $\underline{\mathbf{d}}_{I+1,0}^{m,0}, \ldots, \underline{\mathbf{d}}_{I+1,K+n-1}^{m,0}$ found in this manner, interpret them as a control polygon, and evaluate at v. The resulting point $\underline{\mathbf{d}}_{I+1,J+1}^{m,n}(u,v)$ is the desired projective point on the surface, and is mapped into affine space upon division by the fourth coordinate.

Conversion to Bézier form follows the same recipe: first convert all projective control net rows into Bézier control polygons, following Section 10.5. This gives an intermediate control net, of Bézier form for each row, and of B-spline form for each column. Take each column, interpret it as a B-spline polygon, and convert to Bézier form. The result, after dividing by the fourth coordinate, is the desired piecewise Bézier net together with its weights. Figure 11.11 gives an example.

In order to compute u- and v- partials, we can again compute a bilinear osculant, either from the piecewise Bézier form or from the intermediate points of the de Boor algorithms. In that notation, the four control points of the bilinear osculant are given by

$$\underline{\mathbf{d}}_{I,J}^{m-1,n-1}(u,v), \underline{\mathbf{d}}_{I+1,J}^{m-1,n-1}(u,v), \underline{\mathbf{d}}_{I,J+1}^{m-1,n-1}(u,v), \underline{\mathbf{d}}_{I+1,J+1}^{m-1,n-1}(u,v).$$

11.7 Surfaces of Revolution

Surfaces of revolution play an important role in all of engineering and manufacturing. Using integral surface patches, we can only approximate them – rational surfaces allow an exact representation, provided that the *generatrix* is itself a NURB curve. In general, a generatrix is a curve that sweeps out a surface. In our case, we define an arbitrary straight line $\mathbf{L}$, call it the *axis* of the surface of revolution, and let the generatrix rotate around the axis, thus sweeping out the surface. Clearly, such surfaces are *Euclidean* objects, since they contain infinitely many circles, the *meridians*.

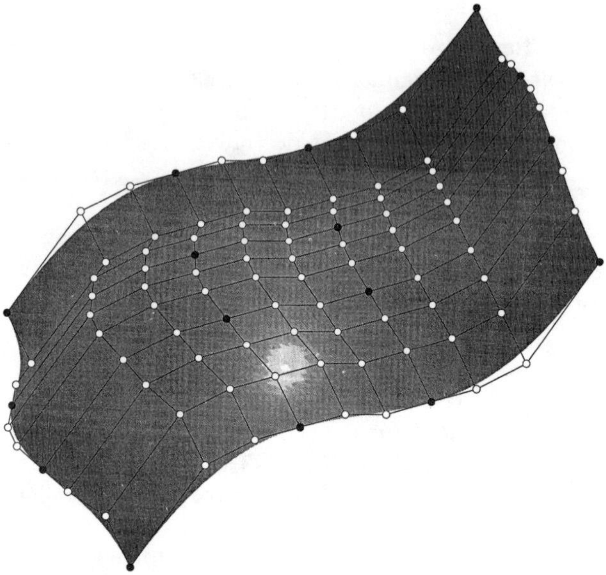

Figure 11.11. Conversion to Bézier form: the surface from the previous figure, but now broken down into rational Bézier patches.

In most practical applications, the generatrix is planar, and lies in a plane also containing the axis.

Analytically, a surface of revolution is given by

$$\mathbf{x}(u,v) = \begin{bmatrix} r(v)\cos u \\ r(v)\sin u \\ z(v) \end{bmatrix}.$$

For fixed v, an isoparametric line $v = const$ traces out a circle of radius $r(v)$, namely a meridian. Since a circle may be exactly represented by rational quadratic arcs, we may find an exact rational representation of a surface of revolution provided we can represent $r(v), z(v)$ in rational form.

The most convenient way to define a surface of revolution is to prescribe the generatrix, given by

$$\mathbf{g}(v) = [r(v), 0, z(v)]^{\mathrm{T}}$$

and by the axis, in the same plane as $\mathbf{g}$. Suppose $\mathbf{g}$ is given by its control polygon, knot sequence, and weight sequence. We can construct a surface of revolution such that each meridian consists of four rational quadratic arcs, as shown in Figure 5.9. For each vertex of the generating polygon, construct a square (perpendicular to the axis of revolution) as in Figure 5.9. Assign the given weights of the generatrix to the four polygons corresponding to the square's edge midpoints; the remaining weights, corresponding to the square's corners, are multiplied by $\cos(45^o) = \sqrt{2}/2$. In this way, we can represent "classical" surfaces such as cylinders, spheres, or tori. See Chapter 15 for numerical examples, and Figure 11.12 for an illustration.

Note that although the generatrix may be defined over a knot sequence $\{v_j\}$ with only simple knots, this is not possible for the knots of the meridian circles: we have to use double knots, thereby essentially reducing the meridians to the piecewise Bézier form. Even after degree elevation to rational cubic form, we must continue to use double knots, as the curve is still not twice differentiable in projective space.

11.8 Developable Surfaces

In Section 7.11, we encountered control plane curves, defined by

$$\underline{\mathbf{B}}(t) = \sum_{i=0}^{n} \underline{\mathbf{B}}_i(t), \tag{11.14}$$

with $\underline{\mathbf{B}}(t)$ as well as the $\underline{\mathbf{B}}_i$ being planes. As t varies, the planes $\underline{\mathbf{B}}(t)$ envelop a surface, a so-called *developable surface* (11.14). Each $\underline{\mathbf{B}}(t)$ is tangent to this surface along a whole line; the collection of all these lines is the envelope of the *edge of regression* of the developable surface. The edge of regression is a sharp ridge on the surface; the surface consists of all tangents of the edge of regression. Figure 11.13 illustrates.

The reason for the name "developable surface" is a physical analogue: such surfaces may be flattened out, or unwrapped, into the plane without any distortions. More algebraically, their Gaussian curvature vanishes everywhere.[7] For these differential geometry concepts, see [25], [47], or [40].

A warning concerning a common misconception: although developable surfaces are ruled, not all ruled surfaces are developable! An affine example is supplied by the bilinear patch: it is ruled (even doubly ruled), but its Gaussian curvature is negative everywhere, and hence it is not developable.

[7] The Gaussian curvature of a surface is a measure for the "roundness" of a surface at a given point. Every point on a hyperboloid has negative Gaussian curvature; every point on an ellipsoid has positive Gaussian curvature, and every point on a cylinder has zero Gaussian curvature.

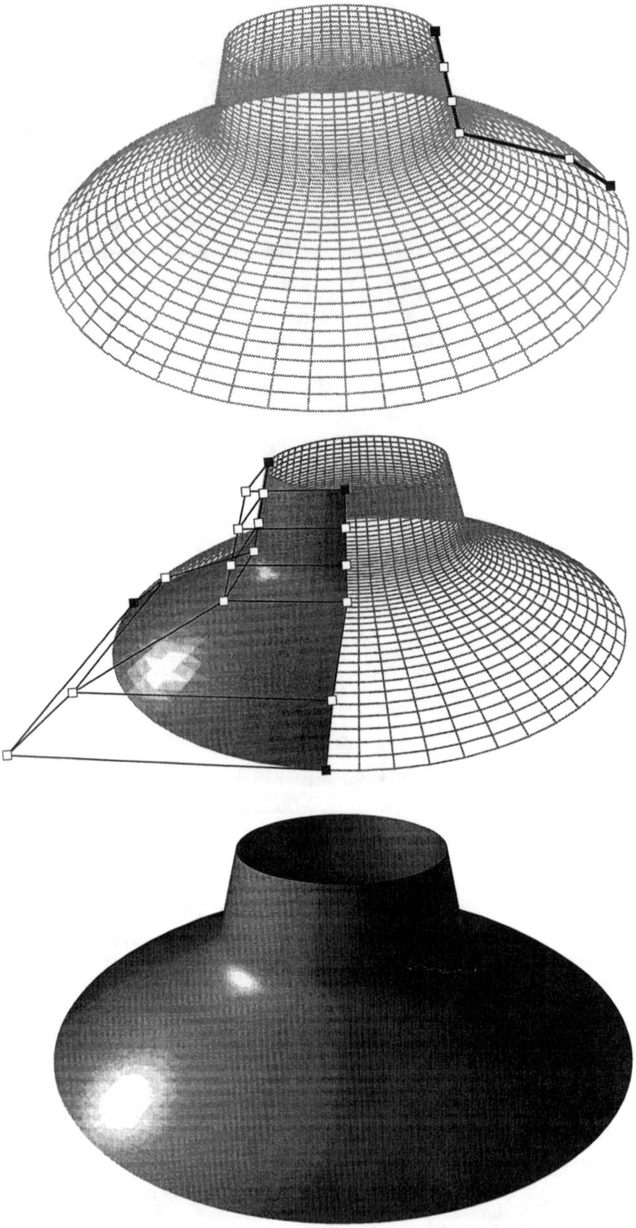

Figure 11.12. Surfaces of revolution: top, the surface with the generatrix control polygon; middle, one quarter of the surface with its control polygon; bottom, shaded view of whole surface.

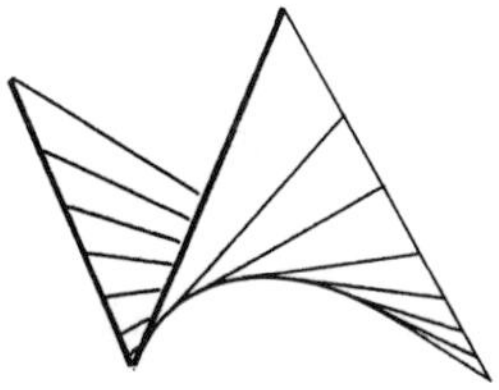

Figure 11.13. Developable surfaces: all tangents to a space curve form a developable surface. The space curve is its edge of regression.

Let $\underline{\mathbf{b}}(t)$ be a cubic that is the edge of regression of a developable surface $\underline{\mathbf{d}}$. At a given parameter value t, all points of the corresponding tangent are given by the blossoms $\underline{\mathbf{b}}[t,t,s]$, i.e., as s traces out all reals, $\underline{\mathbf{b}}[t,t,s]$ traces out all of the tangent to $\underline{\mathbf{b}}(t)$. Thus every point on $\underline{\mathbf{d}}$ can be written as

$$\underline{\mathbf{d}}(s,t) = \underline{\mathbf{b}}[t,t,s]. \tag{11.15}$$

If we think of s as being fixed and t varying, then $\underline{\mathbf{b}}[t,t,s]$ defines a conic; comparing with Section 7.9, we see that it is the cubic's *osculant* at $\underline{\mathbf{b}}(s)$. Thus $\underline{\mathbf{d}}$ does not only contain a family of lines, but a family of conics as well! Each of these conics may be thought of as the intersection of $\underline{\mathbf{d}}$ and one of its *osculating planes*, see Section 7.9.

Let us now consider two of these conics, $\underline{\mathbf{a}}(t)$ and $\underline{\mathbf{c}}(t)$, say. Each of them, illustrated in Figure 11.14, is given by a set of control points $\{\underline{\mathbf{a}}_i\}$ and $\{\underline{\mathbf{c}}\}_i$, respectively, plus a shoulder tangent each (or, alternatively, two weight points each).

Let $\underline{\mathbf{a}}(t)$ and $\underline{\mathbf{c}}(t)$ be the cubic's osculants at $t=\alpha$ and $t=\gamma$, respectively. We can write their control points as blossom values:

$$\underline{\mathbf{a}}_0 = \underline{\mathbf{b}}[0,0,\alpha], \underline{\mathbf{a}}_1 = \underline{\mathbf{b}}[0,1,\alpha], \underline{\mathbf{a}}_2 = \underline{\mathbf{b}}[1,1,\alpha]$$

and

$$\underline{\mathbf{c}}_0 = \underline{\mathbf{b}}[0,0,\gamma], \underline{\mathbf{c}}_1 = \underline{\mathbf{b}}[0,1,\gamma], \underline{\mathbf{c}}_2 = \underline{\mathbf{b}}[1,1,\gamma].$$

Thus the cross ratios formed by four points on the edges of $\underline{\mathbf{b}}$'s control polygon are all equal:

$$\mathrm{cr}(\underline{\mathbf{b}}_0,\underline{\mathbf{b}}_1,\underline{\mathbf{a}}_0,\underline{\mathbf{c}}_0) = \mathrm{cr}(\underline{\mathbf{b}}_1,\underline{\mathbf{b}}_2,\underline{\mathbf{a}}_1,\underline{\mathbf{c}}_1) = \mathrm{cr}(\underline{\mathbf{b}}_2,\underline{\mathbf{b}}_3,\underline{\mathbf{a}}_2,\underline{\mathbf{c}}_2). \tag{11.16}$$

After this investigation of properties of developable surfaces, we now turn to a more practical question: given two arbitrary conic segments $\underline{\mathbf{a}}(t)$ and $\underline{\mathbf{c}}(t)$, can we find a cubic developable surface $\underline{\mathbf{d}}(s,t)$ of the form

$$\underline{\mathbf{d}}(s,t) = (1-s)\underline{\mathbf{a}}(t) + s\underline{\mathbf{c}}(t)? \tag{11.17}$$

In general, the answer is negative, as Figure 11.14 illustrates: we would have to find $\underline{\mathbf{d}}$'s edge of regression $\underline{\mathbf{b}}(t)$. Then the four points $\underline{\mathbf{a}}_0, \underline{\mathbf{c}}_0, \underline{\mathbf{a}}_1, \underline{\mathbf{c}}_1$ have to be coplanar, namely they have to be in $\underline{\mathbf{b}}$'s osculating plane at $\underline{\mathbf{b}}_0$, spanned by $\underline{\mathbf{b}}_0, \underline{\mathbf{b}}_1, \underline{\mathbf{b}}_2$. Similarly, the four points $\underline{\mathbf{a}}_1, \underline{\mathbf{c}}_1, \underline{\mathbf{a}}_2, \underline{\mathbf{c}}_2$ must lie in $\underline{\mathbf{b}}$'s osculating plane at $\underline{\mathbf{b}}_3$.

Suppose then that our two conics meet these coplanarity conditions. Our first aim is to find the control polygon for the edge of regression. We immediately find $\underline{\mathbf{b}}_1$ as the intersection of the lines through $\underline{\mathbf{a}}_0, \underline{\mathbf{c}}_0$ and $\underline{\mathbf{a}}_1, \underline{\mathbf{c}}_1$ and $\underline{\mathbf{b}}_2$ as the intersection of the two lines through $\underline{\mathbf{a}}_1, \underline{\mathbf{c}}_1$ and $\underline{\mathbf{a}}_2, \underline{\mathbf{c}}_2$.

Next, we compute the cross ratio formed by the four points $\underline{\mathbf{b}}_1, \underline{\mathbf{b}}_2, \underline{\mathbf{a}}_1, \underline{\mathbf{c}}_1$. We can then find $\underline{\mathbf{b}}_0$ and $\underline{\mathbf{b}}_3$ from the cross ratio condition (11.16), thus having found the desired control polygon.

We have not yet determined the weight points of the two conics. For this task, recall that two conic osculants to a cubic must share a common tangent, as shown in Figure 7.9. This tangent must then be the intersection of the two planes that contain the conics, a condition that will not be met by two arbitrary conics. If $\underline{\mathbf{q}}_i^a$ and $\underline{\mathbf{q}}_i^c$ are the weight points of the two conics, and $\underline{\mathbf{q}}_i$ are those of the cubic, it is not hard to show that $\underline{\mathbf{q}}_0, \underline{\mathbf{q}}_1, \underline{\mathbf{q}}_0^a, \underline{\mathbf{q}}_0^c$ are collinear. Similarly, $\underline{\mathbf{q}}_1, \underline{\mathbf{q}}_2, \underline{\mathbf{q}}_1^a, \underline{\mathbf{q}}_1^c$ are collinear. Thus we can prescribe $\underline{\mathbf{q}}_0^a$ and $\underline{\mathbf{q}}_0^c$ and determine the remaining $\underline{\mathbf{q}}_1^a, \underline{\mathbf{q}}_1^c$.

This construction relies heavily upon properties that the given conics must satisfy. In real applications, this will rarely be the case, and approximative solutions must be sought, as outlined in [50].

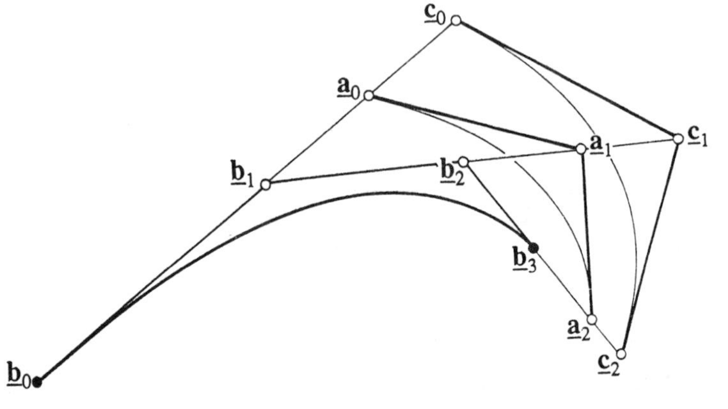

Figure 11.14. Conics on developables: two conic segments together with the edge of regression.

Developable surfaces play an important role in several applications. In the stamping process of sheet metals, so-called *binder surfaces* are used to clamp the sheet metal around the die. Initially, the sheet falls onto the binder surface without touching the die. In order to touch the binder surface everywhere without wrinkling, the binder must be developable.[8]

Another application is in the shipbuilding industry, where ship hulls are assembled from rolled sheets — again forming developable surfaces.

More literature on this topic: Aumann [7], Ravani and Boddulori [19], Frey [17], Pottmann and Farin [50], Redont [98].

11.9 Problems

1. Show that the diagonal curves of a rational bilinear patch cannot be both elliptic or both hyperbolic.

2. Verify the following:
   a) A hyperboloid of one sheet is obtained as a surface of revolution with the hyperbola as the generatrix and the axis of symmetry between the two branches as axis.
   b) For a hyperboloid of two sheets, the generatrix is again the hyperbola, but the axis is the common axis of symmetry of both branches.

3. Show that the planar weight point quadrilaterals from Section 11.4 are parallelograms if all weights are equal. Note that this is strictly an affine phenomenon.

4. Show that every planar section of a bilinear patch in $\mathbb{P}^3$ is a conic section.

[8]In practice, it suffices that the binder is close to being developable.

12

Rational Bézier Triangles

Bézier triangles, short for triangular Bézier patches, actually owe their discovery to P. de Casteljau from Citröen, who outlined their theory in the two technical reports [38] and [39]. Here, we are interested in their rational counterparts, but for the sake of simplicity, we start with an outline of integral, polynomial Bézier triangles in projective space. These will then be projected onto their rational counterparts.

Triangular patches enjoy some popularity among CAGD researchers, although they have not been included in standards such as IGES. They are useful when it comes to modeling complex shapes that do not lend themselves to a rectangular topology. *Rational* Bézier triangles have not been investigated very thoroughly yet; some pointers to the literature include [24].

12.1 Projective Bézier Triangles

Before exploring rational Bézier triangles, we introduce their projective preimages – triangular polynomial patches in $\mathbb{P}^3$, expressed in Bernstein-Bézier form. They behave much like their integral (polynomial) counterparts. For more complete information about integral Bézier triangles, the reader is referred to [44] or [47].

A Bézier patch in $I\!P^3$ is defined by a control net similar to those of tensor product patches, except that it comes in a triangular arrangement. It is convenient to give each control point three subscripts, and a typical control net then looks like this:

$$\begin{array}{c} \underline{\mathbf{b}}_{040} \\ \underline{\mathbf{b}}_{031}\underline{\mathbf{b}}_{130} \\ \underline{\mathbf{b}}_{022}\underline{\mathbf{b}}_{121}\underline{\mathbf{b}}_{220} \\ \underline{\mathbf{b}}_{013}\underline{\mathbf{b}}_{112}\underline{\mathbf{b}}_{211}\underline{\mathbf{b}}_{310} \\ \underline{\mathbf{b}}_{004}\underline{\mathbf{b}}_{103}\underline{\mathbf{b}}_{202}\underline{\mathbf{b}}_{301}\underline{\mathbf{b}}_{400} \end{array}$$

Note how all subscripts sum to four – this indicates that we are dealing with the control net of a degree four, or quartic, patch. For other degrees n, the arrangement is similar, and the control point subscripts always sum to n. The control points $\underline{\mathbf{b}}_{\mathbf{i}} = \underline{\mathbf{b}}_{i,j,k}$ are points in $I\!P^3$, thus having four coordinates x, y, z, w each.

Consider an arbitrary (nondegenerate) triangle in the projective plane, which, together with a fourth point, defines a projective reference frame and thus projective coordinates $\underline{\mathbf{u}} = [u, v, w]^{\mathrm{T}}$.[1] We give the triangle vertices the coordinates $\mathbf{e1} = [1,0,0], \mathbf{e2} = [0,1,0], \mathbf{e3} = [0,0,1]$, and the fourth point the coordinates $[1/3, 1/3, 1/3]$. A triangular patch $\underline{\mathbf{b}}(\underline{\mathbf{u}})$ is a polynomial map of the projective plane into $I\!P^3$, such that a point $\underline{\mathbf{x}}$ on the triangular patch is expressed as

$$\underline{\mathbf{b}}^n(\underline{\mathbf{u}}) = \underline{\mathbf{b}}_{\mathbf{0}}^n(\underline{\mathbf{u}}) = \sum_{|\mathbf{j}|=n} \underline{\mathbf{b}}_{\mathbf{j}} B_{\mathbf{j}}^n(\underline{\mathbf{u}}), \tag{12.1}$$

where the $B_{\mathbf{i}}^n(\underline{\mathbf{u}})$ are trivariate[2] Bernstein polynomials:

$$B_{\mathbf{i}}^n(\underline{\mathbf{u}}) = B_{i,j,k}^n(u,v,w) = \frac{n!}{i!j!k!} u^i v^j w^k; \quad |\mathbf{i}| = n. \tag{12.2}$$

Figure 12.1 gives an example of a cubic patch with its control net.

Bernstein polynomials satisfy the following recursion:

$$B_{\mathbf{i}}^n(\underline{\mathbf{u}}) = uB_{\mathbf{i}-\mathbf{e1}}^{n-1}(\underline{\mathbf{u}}) + vB_{\mathbf{i}-\mathbf{e2}}^{n-1}(\underline{\mathbf{u}}) + wB_{\mathbf{i}-\mathbf{e3}}^{n-1}(\underline{\mathbf{u}}); \quad |\mathbf{i}| = n, \tag{12.3}$$

which can be used to define the de Casteljau algorithm for the evaluation of Bézier triangles:

[1] The third projective coordinate, w, for $I\!P^2$ is named the same as the fourth coordinate for $I\!P^3$. As long as there is no danger of confusion, we will use this slightly ambiguous notation.

[2] Note that we are still dealing with *surfaces*, since the u, v, w are coordinates of a point in the plane!

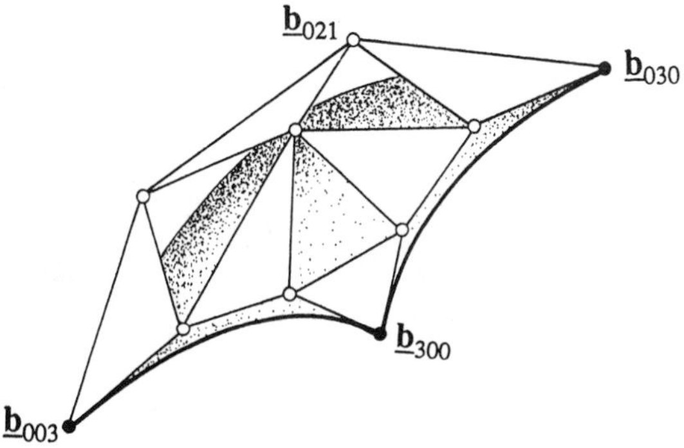

Figure 12.1. Bézier triangles: a cubic patch with its control net.

de Casteljau algorithm:

Given: A triangular array of points $\underline{\mathbf{b}}_{\mathbf{i}} \in \mathbb{P}^3$; $|\mathbf{i}| = n$ and a point in $\mathbb{P}^3$ with coordinates $\underline{\mathbf{u}}$.

Set:

$$\underline{\mathbf{b}}_{\mathbf{i}}^r(\underline{\mathbf{u}}) = u\underline{\mathbf{b}}_{\mathbf{i+e1}}^{r-1}(\underline{\mathbf{u}}) + v\underline{\mathbf{b}}_{\mathbf{i+e2}}^{r-1}(\underline{\mathbf{u}}) + w\underline{\mathbf{b}}_{\mathbf{i+e3}}^{r-1}(\underline{\mathbf{u}}) \tag{12.4}$$

where

$$r = 1, \ldots, n \quad \text{and} \quad |\mathbf{i}| = n - r$$

and $\underline{\mathbf{b}}_{\mathbf{i}}^0(\underline{\mathbf{u}}) = \underline{\mathbf{b}}_{\mathbf{i}}$. Then $\underline{\mathbf{b}}_{\mathbf{0}}^n(\underline{\mathbf{u}})$ is the point with parameter value $\underline{\mathbf{u}}$ on the *Bézier triangle* $\underline{\mathbf{b}}^n$.

Figure 12.2 shows the quadratic case.

This algorithm is a direct generalization of the univariate de Casteljau algorithm, and shares many properties with it. In particular, the three intermediate points $\underline{\mathbf{b}}_{\mathbf{e1}}^{n-1}, \underline{\mathbf{b}}_{\mathbf{e2}}^{n-1}, \underline{\mathbf{b}}_{\mathbf{e3}}^{n-1}$, span the tangent plane of the patch at $\underline{\mathbf{b}}(\underline{\mathbf{u}})$.

The boundary curves of a triangular patch are determined by the boundary control vertices (having at least one zero as a subscript). For example, a point on the boundary curve $\underline{\mathbf{b}}^n(u, 0, w)$ is generated by

$$\underline{\mathbf{b}}_{\mathbf{i}}^r(u, 0, w) = u\underline{\mathbf{b}}_{\mathbf{i+e1}}^{r-1} + w\underline{\mathbf{b}}_{\mathbf{i+e3}}^{r-1},$$

which is the univariate de Casteljau algorithm for Bézier curves.

In analogy to the weight points for Bézier curves, we may define

$$\underline{\mathbf{q}}_{\mathbf{i}} = \underline{\mathbf{b}}_{\mathbf{i+e1}} + \underline{\mathbf{b}}_{\mathbf{i+e2}} + \underline{\mathbf{b}}_{\mathbf{i+e3}}; \quad |\mathbf{i}| = n - 1.$$

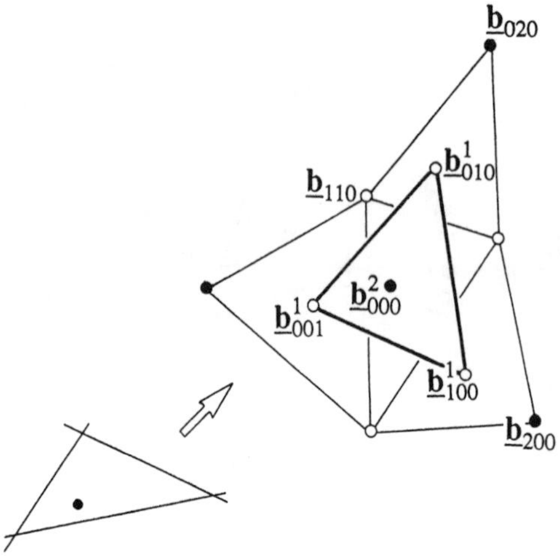

Figure 12.2. The de Casteljau algorithm: the quadratic case. The tangent plane at the computed point is highlighted.

These points are only defined in the "upright" subtriangles;, see Figure 12.3 for an illustration. In the curve case, we could give each of the weight points an arbitrary position on its control polygon leg; this is not possible anymore: changing just one of the weight points would result in conflicts with neighboring "upside down" subtriangles.

G. Albrecht has resolved this problem: if the weight points are assigned in a spiral pattern, no conflicts arise. See Figure 12.4 and [6].

12.2 Degree Elevation

It is possible to write $\underline{\mathbf{b}}^n$ as a Bézier triangle of degree $n+1$:

$$\sum_{|\mathbf{i}|=n} \underline{\mathbf{b}}_{\mathbf{i}} B_{\mathbf{i}}^n(\mathbf{u}) = \sum_{|\mathbf{i}|=n+1} \underline{\mathbf{b}}_{\mathbf{i}}^{(1)} B_{\mathbf{i}}^{n+1}(\mathbf{u}). \tag{12.5}$$

The control points $\underline{\mathbf{b}}_{\mathbf{i}}^{(1)}$ are obtained from

$$\underline{\mathbf{b}}_{\mathbf{i}}^{(1)} = \frac{1}{n+1}[i\underline{\mathbf{b}}_{\mathbf{i}-\mathbf{e1}} + j\underline{\mathbf{b}}_{\mathbf{i}-\mathbf{e2}} + k\underline{\mathbf{b}}_{\mathbf{i}-\mathbf{e3}}]. \tag{12.6}$$

For a proof, we multiply the left-hand side of (12.5) by $u+v+w$ and compare coefficients of like powers. Figure 12.5 illustrates the case $n=2$.

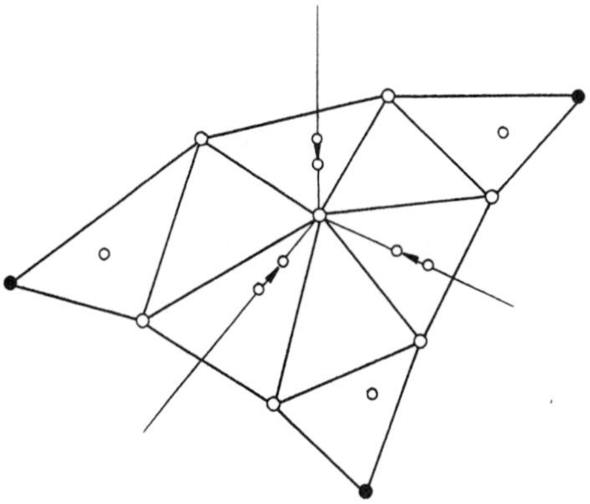

Figure 12.3. Weight points: these may be defined in all "upright" subtriangles. If we increase the center weight, three weight points will move towards it.

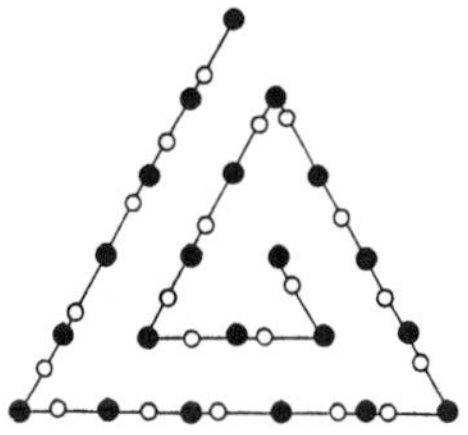

Figure 12.4. Weight points: a spiral pattern yields to a consistent patch definition. Solid: Bézier points; open: weight points.

12.3 Blossoms and Derivatives

The *blossom* of a triangular patch is obtained by feeding n possibly different arguments into the de Casteljau algorithm; it is denoted by $\underline{\mathbf{b}}[\underline{\mathbf{u}}_1, \ldots, \underline{\mathbf{u}}_n]$. Note that each argument consists of three components, denoting a point in $\mathbb{P}^2$. If all n arguments agree, we obtain the point on the surface corresponding to that argument.

Now let $\underline{\mathbf{a}}, \underline{\mathbf{b}}, \underline{\mathbf{c}}$ be a second triangle in $\mathbb{P}^2$. Our polynomial $\underline{\mathbf{b}}(\mathbf{u})$ is defined over all of $\mathbb{P}^2$, and we may ask for its control points relative to the

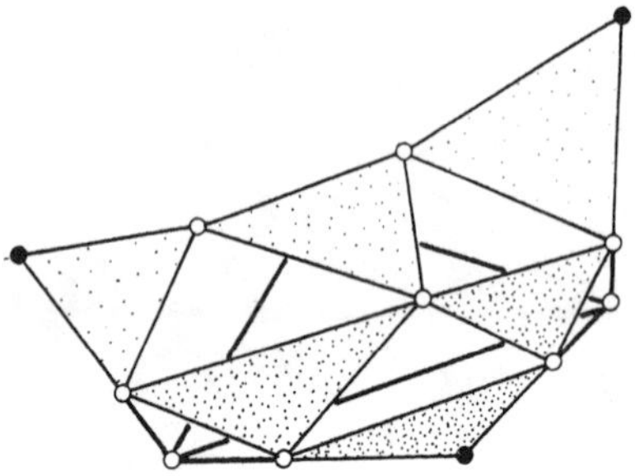

Figure 12.5. Degree elevation: a quadratic control net together with the equivalent cubic net (shaded).

triangle $\underline{\mathbf{a}}, \underline{\mathbf{b}}, \underline{\mathbf{c}}$. They are given simply by

$$\underline{\mathbf{c}}_{\mathbf{i}} = \underline{\mathbf{b}}[\underline{\mathbf{a}}^{<i>}, \underline{\mathbf{b}}^{<j>}, \underline{\mathbf{c}}^{<k>}]; \quad |\mathbf{i}| = n. \tag{12.7}$$

We may use this approach to compute the control points of a line on the surface: let $\underline{\mathbf{a}}$ and $\underline{\mathbf{b}}$ be two points in $\mathbb{P}^2$. They define a line, which will be mapped to a polynomial curve of degree n, lying completely on the surface $\underline{\mathbf{b}}(\underline{\mathbf{u}})$. The control points $\underline{\mathbf{p}}_i$ of this curve are simply

$$\underline{\mathbf{p}}_i = \underline{\mathbf{b}}[\underline{\mathbf{a}}^{<n-i>}, \underline{\mathbf{b}}^{<i>}]; \quad i = 0, \ldots, n. \tag{12.8}$$

Note the difference to rectangular patches: there a line in the domain maps to a curve of degree $m+n$ in space!

Our projective reference frame for $\mathbb{P}^2$ implies the existence of a *fundamental line* $\underline{\mathbf{F}}$; see Section 1.5. Let $\underline{\mathbf{f}}$ be a point on it, whereas the vertices of the domain triangle are not on it. In complete analogy to our development of the curve case, we may use $\underline{\mathbf{f}}$ to compute *derivatives* of $\underline{\mathbf{b}}(\underline{\mathbf{u}})$ with respect to $\underline{\mathbf{f}}$:

$$D^r_{\underline{\mathbf{f}}}\underline{\mathbf{b}}(\underline{\mathbf{u}}) \hat{=} \underline{\mathbf{b}}[\underline{\mathbf{f}}^{<r>}, \underline{\mathbf{u}}^{<n-r>}]. \tag{12.9}$$

If we set $r = n$, then we see that the nth derivative of $\underline{\mathbf{b}}$ with respect to $\underline{\mathbf{f}}$ is on the image of $\underline{\mathbf{F}}$, which is a curve on the surface.

We shall consider the case $r = 1$ in more detail. Setting $\underline{\mathbf{f}} = [f, g, h]^{\mathrm{T}}$, we have

$$\begin{aligned} D^r_{\underline{\mathbf{f}}}\underline{\mathbf{b}}(\underline{\mathbf{u}}) &\hat{=} \underline{\mathbf{b}}[\underline{\mathbf{f}}^{<r>}, \underline{\mathbf{u}}^{<n-r>}] \\ &\hat{=} \sum_{|\mathbf{i}|=n-1} [f\underline{\mathbf{b}}_{\mathbf{i}+\mathbf{e1}} + g\underline{\mathbf{b}}_{\mathbf{i}+\mathbf{e2}} + h\underline{\mathbf{b}}_{\mathbf{i}+\mathbf{e3}}]B^{n-1}_{\mathbf{i}}(\underline{\mathbf{u}}). \end{aligned}$$

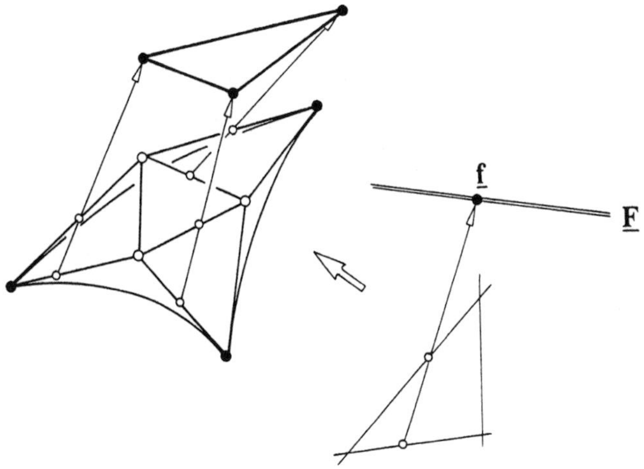

Figure 12.6. Derivatives: the derivative with respect to a point on the fundamental line in the domain is given by a surface whose control points are obtained by the blossoming principle.

Thus the coefficients of the derivative surface (of degree $n-1$) are formed by the terms $f\underline{\mathbf{b}}_{\mathbf{i+e1}}+g\underline{\mathbf{b}}_{\mathbf{i+e2}}+h\underline{\mathbf{b}}_{\mathbf{i+e3}}$, obtained by application of one level of the de Casteljau algorithm to the original control points. Figure 12.6 illustrates.

If $\underline{\mathbf{f}}$ and $\underline{\mathbf{g}}$ are two points on $\underline{\mathbf{F}}$, we may define *mixed derivatives* in the same way:

$$D^{r,s}_{\underline{\mathbf{f}},\underline{\mathbf{g}}}\underline{\mathbf{b}}(\underline{\mathbf{u}})\hat{=}\underline{\mathbf{b}}[\underline{\mathbf{f}}^{<r>},\underline{\mathbf{g}}^{<s>},\underline{\mathbf{u}}^{<n-r-s>}]. \tag{12.10}$$

Note that it is not meaningful to consider expressions of the form $D^{r,s,t}_{\underline{\mathbf{f}},\underline{\mathbf{g}},\underline{\mathbf{h}}}\underline{\mathbf{b}}(\underline{\mathbf{u}})$ with all three points $\underline{\mathbf{f}},\underline{\mathbf{g}},\underline{\mathbf{h}}$ on $\underline{\mathbf{F}}$: this leads to a new domain triangle with three collinear vertices.

Let us investigate the relationship between blossoms and derivatives in more detail for the quadratic case $n=2$. The fundamental line $\underline{\mathbf{F}}$ is mapped to a conic $\underline{\mathbf{c}}_f$ on the surface. Let $\underline{\mathbf{u}}$ be a point in the domain, and $\underline{\mathbf{b}}(\underline{\mathbf{u}})$ its corresponding surface point. With $\underline{\mathbf{f}}$ a point on $\underline{\mathbf{F}}$, the line $\underline{\mathbf{u}}\wedge\underline{\mathbf{f}}$ is mapped to a conic $\underline{\mathbf{c}}$ on the surface. The derivative of this conic at $\underline{\mathbf{u}}$ is the intersection of the conic's tangents at $\underline{\mathbf{b}}(\underline{\mathbf{u}})$ and at $\underline{\mathbf{b}}(\underline{\mathbf{f}})$.[3] Thus every derivative to $\underline{\mathbf{b}}$ lies on one of $\underline{\mathbf{b}}$'s tangent planes at $\underline{\mathbf{c}}_f$. Note that these tangent planes generate a developable surface; see Section 11.8.

[3]Recall that the derivative of a conic at a given point is the intersection with the tangent at that point with the tangent at $t=\infty$.

In the degenerate case where $\underline{\mathbf{c}}_f$ collapses to a single point $\underline{\mathbf{p}}$, all tangent planes along $\underline{\mathbf{c}}$ degenerate to the tangent plane at $\underline{\mathbf{p}}$, and all first derivatives of $\underline{\mathbf{b}}$ lie in this plane. Quadratic surfaces for which $\underline{\mathbf{F}}$ is mapped to a single point $\underline{\mathbf{p}}$ are obtainable by a *stereographic projection* through $\underline{\mathbf{p}}$; see Section 13.3. In this special case, the quadratic surface is actually part of a *quadric*; see [24], [113], [84].

12.4 Barycentric Coordinates

In the projective plane, any four points $\underline{\mathbf{a}}, \underline{\mathbf{b}}, \underline{\mathbf{c}}, \underline{\mathbf{d}}$ define a coordinate system by setting $\underline{\mathbf{d}} = \underline{\mathbf{a}} + \underline{\mathbf{b}} + \underline{\mathbf{c}}$. We can then write any other point $\underline{\mathbf{x}}$ in $\mathbb{P}^2$ as $\underline{\mathbf{x}} = u\underline{\mathbf{a}} + v\underline{\mathbf{b}} + w\underline{\mathbf{c}}$. This representation is unique up to a common factor in the $\underline{\mathbf{x}}$'s coordinates u, v, w. See Section 1.5 for more details.

Since $(u, v, w) = (0, 0, 0)$ does not denote a point location, we may normalize our projective coordinates such that

$$u + v + w = 1. \tag{12.11}$$

Projective coordinates that satisfy (12.11) are called *barycentric coordinates* after F. Moebius [83]. Their main importance is not in projective, but in affine geometry.

If we project the relation $\underline{\mathbf{d}} = \underline{\mathbf{a}} + \underline{\mathbf{b}} + \underline{\mathbf{c}}$ into affine space, we obtain

$$\mathbf{d} = \frac{1}{3}\mathbf{a} + \frac{1}{3}\mathbf{b} + \frac{1}{3}\mathbf{c},$$

as was outlined in Section 1.9. This has the geometric interpretation that $\mathbf{d}$ is the centroid of the triangle $\underline{\mathbf{a}}, \underline{\mathbf{b}}, \underline{\mathbf{c}}$.

Just as in the projective plane, we write every point $\mathbf{x}$ of the extended affine plane as a *barycentric combination*

$$\mathbf{x} = u\mathbf{a} + v\mathbf{b} + w\mathbf{c}; \quad u + v + w = 1.$$

Figure 12.7 shows the three families of constant parameter values, in complete analogy to Figure 1.9. Note that now the three families of constant parameter values each contain parallel lines. Thus the centers of the corresponding pencils all lie on the line at infinity, and hence are collinear. Barycentric coordinates are special projective coordinates: the fundamental line (see Section 1.5) is the line at infinity.

If a point $\mathbf{p}$ is given and we seek its barycentric coordinates $\mathbf{u} = (u, v, w)$ with respect to an affine triangle $\mathbf{a}, \mathbf{b}, \mathbf{c}$, we find

$$u = \frac{\text{area}(\mathbf{p}, \mathbf{b}, \mathbf{c})}{\text{area}(\mathbf{a}, \mathbf{b}, \mathbf{c})}, \quad v = \frac{\text{area}(\mathbf{a}, \mathbf{p}, \mathbf{c})}{\text{area}(\mathbf{a}, \mathbf{b}, \mathbf{c})}, \quad w = \frac{\text{area}(\mathbf{a}, \mathbf{b}, \mathbf{p})}{\text{area}(\mathbf{a}, \mathbf{b}, \mathbf{c})}, \tag{12.12}$$

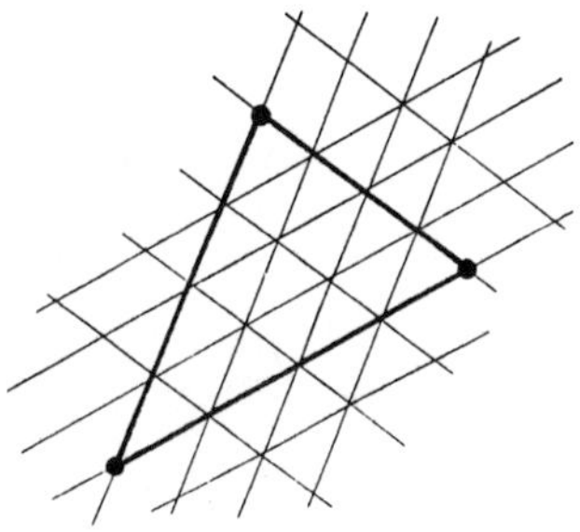

Figure 12.7. Barycentric coordinate systems: the three families of isoparametric lines are shown.

with the understanding that the area of a triangle has a sign! The area of an affine triangle is given by area($\mathbf{a}, \mathbf{b}, \mathbf{c}$) = det($\mathbf{a} - \mathbf{c}, \mathbf{b} - \mathbf{c}$)/2. Here, each point has *two* coordinates!

12.5 Rational Bézier Triangles

A rational Bézier triangle is the projection of a Bézier triangle in $\mathbb{P}^3$ into $\mathbb{E}^3$. We thus have:

$$\mathbf{b}^n(\mathbf{u}) = \mathbf{b}_{\mathbf{0}}^n(\mathbf{u}) = \frac{\sum_{|\mathbf{i}|=n} w_{\mathbf{i}} \mathbf{b}_{\mathbf{i}} B_{\mathbf{i}}^n(\mathbf{u})}{\sum_{|\mathbf{i}|=n} w_{\mathbf{i}} B_{\mathbf{i}}^n(\mathbf{u})}, \tag{12.13}$$

where, as usual, the $w_{\mathbf{i}}$ are the weights associated with the control vertices $\mathbf{b}_{\mathbf{i}}$. Needless to say, for positive weights we have the convex hull property, and we have affine and projective invariance.

Rational Bézier triangles may be evaluated by a de Casteljau algorithm in a not too surprising way:

Rational de Casteljau algorithm

Given: A triangular array of points $\mathbf{b}_{\mathbf{i}} \in \mathbb{E}^3$; $|\mathbf{i}| = n$, corresponding weights $w_{\mathbf{i}}$, and a point in a domain triangle with barycentric coordinates $\mathbf{u}$.

Set:

$$\mathbf{b}_{\mathbf{i}}^r(\mathbf{u}) = \frac{u w_{\mathbf{i+e1}}^{r-1} \mathbf{b}_{\mathbf{i+e1}}^{r-1} + v w_{\mathbf{i+e2}}^{r-1} \mathbf{b}_{\mathbf{i+e2}}^{r-1} + w w_{\mathbf{i+e3}}^{r-1} \mathbf{b}_{\mathbf{i+e3}}^{r-1}}{w_{\mathbf{i}}^r} \tag{12.14}$$

where

$$w_{\mathbf{i}}^r = w_{\mathbf{i}}^r(\mathbf{u}) = uw_{\mathbf{i+e1}}^{r-1}(\mathbf{u}) + vw_{\mathbf{i+e2}}^{r-1}(\mathbf{u}) + ww_{\mathbf{i+e3}}^{r-1}(\mathbf{u})$$

and

$$r = 1, \ldots, n \quad \text{and} \quad |\mathbf{i}| = n - r$$

and $\mathbf{b}_{\mathbf{i}}^0(\mathbf{u}) = \mathbf{b}_{\mathbf{i}}, w_{\mathbf{i}}^0 = w_{\mathbf{i}}$. Then $\mathbf{b}_{\mathbf{0}}^n(\mathbf{u})$ is the point with parameter value $\mathbf{u}$ on the rational Bézier triangle $\mathbf{b}^n$.

This algorithm can be interpreted as projecting every intermediate point $\underline{\mathbf{b}}_i^r$ into affine space. Of course, we could evaluate all four components of $\underline{\mathbf{b}}(\mathbf{u})$ separately and divide through by the w-component as a final step.

Here is a formula for the *directional derivative* of a rational Bézier triangle. Let $\mathbf{d}$ denote a direction in the domain triangle, expressed in barycentric coordinates. This means that $\mathbf{d}$ is the difference between two points, and thus a vector. Note that $\mathbf{d}$'s barycentric coordinates sum to zero. We are interested in the directional derivative $D_{\mathbf{d}}$ of a rational triangular Bézier patch $\mathbf{b}^n(\mathbf{u})$. Proceeding exactly as in the curve case (7.28), we obtain:

$$D_{\mathbf{d}}\mathbf{b}^n(\mathbf{u}) = \frac{1}{w(\mathbf{u})}[\dot{\mathbf{p}}(\mathbf{u}) - D_{\mathbf{d}}(\mathbf{u})\mathbf{b}^n(\mathbf{u})],$$

where we have set

$$\mathbf{p}(\mathbf{u}) = w(\mathbf{u})\mathbf{b}^n(\mathbf{u}) = \sum_{|\mathbf{i}|=n} w_{\mathbf{i}}\mathbf{b}_{\mathbf{i}}B_{\mathbf{i}}^n(\mathbf{u}).$$

Higher derivatives follow the pattern of (7.31):

$$D_{\mathbf{d}}^r\mathbf{b}^n(\mathbf{u}) = \frac{1}{w(\mathbf{u})}\left[D_{\mathbf{d}}^r\mathbf{p}(\mathbf{u}) - \sum_{j=1}^{r} D_{\mathbf{d}}^j w(\mathbf{u}) D_{\mathbf{d}}^{r-j}(\mathbf{u})\right].$$

Similar to the rectangular case in Section 11.3, we may use the intermediate points in the de Casteljau algorithm to determine the *tangent plane* corresponding to a parameter value $\mathbf{u}$: it is given by the three $\mathbf{b}_{\mathbf{e1}}^{n-1}(\mathbf{u}), \mathbf{b}_{\mathbf{e2}}^{n-1}(\mathbf{u}), \mathbf{b}_{\mathbf{e3}}^{n-1}(\mathbf{u})$.

An example of a rational Bézier triangle is given in Section 15.6; there, an octant of a sphere is described.

12.6 Problems

1. The isoparametric lines $u = const, v = const, w = const$ of a rational quadratic Bézier triangle are conics. At a given point, we can thus define three conics through it. Can all of these conics be of the same type?

2. What is the minimum degree for a rational Bézier triangle such that it has *two* ellipses as the intersection with a plane?

3. What is the minimum degree for a rational Bézier triangle such that it can define a surface with the topology of a torus?

13

Quadrics

The most fundamental curves in the projective plane are the conics. As it turns out, they can be represented by projective quadratic Bézier curves. Conversely, every projective quadratic Bézier curve represents a conic.

For surfaces, life is not that easy. The fundamental surfaces in projective space are the *quadrics* — but there is no one-to-one correspondence between them and any kind of Bézier surface. In this chapter, we shall define quadrics, and then explore the relationship between quadrics and Bézier surfaces.

13.1 Quadrics

A quadric in $\mathbb{P}^3$ is a surface that has an implicit equation of the form

$$\underline{\mathbf{x}}^{\mathrm{T}} A \underline{\mathbf{x}} = 0, \tag{13.1}$$

with $\underline{\mathbf{x}} = [x, y, z, w]^{\mathrm{T}}$. A quadric has no more than two intersections with any straight line. Any planar section of a quadric is a conic. In fact, an alternative definition of a quadric is as being a surface all of whose planar sections are conics.

A quadric is defined by nine points on it. The implicit form takes on the form

$$\begin{vmatrix} x^2 & y^2 & z^2 & w^2 & xy & xz & yz & xw & yw & zw \\ x_1^2 & y_1^2 & z_1^2 & w_1^2 & x_1y_1 & x_1z_1 & y_1z_1 & x_1w_1 & y_1w_1 & z_1w_1 \\ \vdots & \vdots & \vdots & \vdots & \vdots & \vdots & \vdots & \vdots & \vdots & \vdots \\ x_9^2 & y_9^2 & z_9^2 & w_9^2 & x_9y_9 & x_9z_9 & y_9z_9 & x_9w_9 & y_9w_9 & z_9w_9 \end{vmatrix} = 0, \tag{13.2}$$

much in the same way as (4.25) described a conic through five points.

Let $\underline{\mathbf{a}}, \underline{\mathbf{b}}, \underline{\mathbf{c}}$ be any three noncollinear points on a quadric. They form a plane $\underline{\mathbf{P}}$, which cuts a conic Γ out of the quadric. The tangent planes at $\underline{\mathbf{a}}, \underline{\mathbf{b}}, \underline{\mathbf{c}}$ cut tangents to Γ out of $\underline{\mathbf{P}}$; see Figure 13.1. The three points $\underline{\mathbf{a}}, \underline{\mathbf{b}}, \underline{\mathbf{c}}$ together with the tangents, all belonging to Γ, must then satisfy Brianchon's theorem (see Figure 3.13): the three heavy lines in Figure 13.1 are concurrent. This shows that three tangent planes to a quadric are not independent of each other. In particular, a quadric is not determined by three points and three tangent planes, although these might naively be counted as nine points.

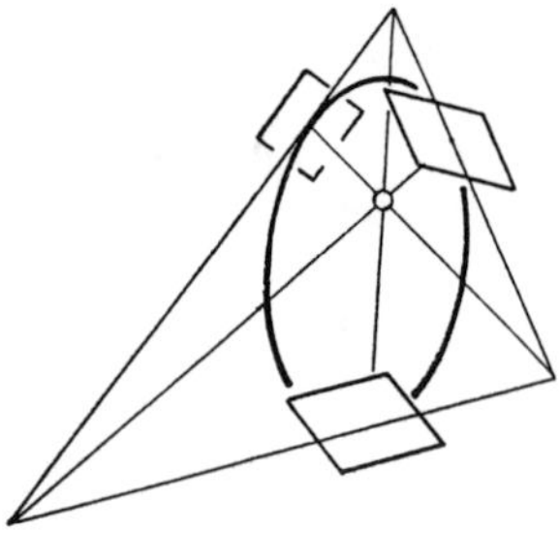

Figure 13.1. Tangent planes to a quadric: the shown configuration must satisfy Brianchon's theorem.

13.2 Quadric Classification

All quadrics in $\mathbb{P}^3$ can be classified into three categories:[1]

a) If for every point on the quadric, the tangent plane intersects the quadric in two distinct straight lines, the quadric is called doubly ruled or annular.

[1] We limit ourselves to *real* quadrics here – if we allow *complex* quadrics, the distinction between the three quadric types disappears.

b) If for every point on the quadric, the tangent plane intersects the quadric in exactly one straight line, the quadric is called singly ruled or a cone.

c) If for every point on the quadric, the tangent plane does not intersect the quadric in any straight line, the quadric is called oval.

Since straight lines are mapped to straight lines by projective maps, it follows that quadrics can only be mapped to quadrics in the same category. Figure 13.2 shows the local geometry of the three categories. This is in sharp contrast to the curve case: there, every conic could be obtained as a projective map of any other conic. Now, we have to pay attention to the kind of quadric with which we are dealing.

Affine quadrics are obtained by projecting the projective quadrics into affine space. Obviously, the category of a quadric is not changed by this projection. When we discussed affine conics, we studied their behavior at the line at infinity. Consequently, we are now interested in the behavior of quadrics at the plane at infinity. Let $\underline{\mathbf{H}}$ be the projective plane that will be mapped to the affine plane at infinity. We will call affine quadrics

- *hyperboloids* if they intersect $\underline{\mathbf{H}}$,
- *ellipsoids* have no point in common with $\underline{\mathbf{H}}$,
- *paraboloids* if they are tangent to $\underline{\mathbf{H}}$,

in complete analogy to the conic classification from Section 3.5. If the projective quadric intersects $\underline{\mathbf{H}}$, the resulting conic will be mapped to an affine *conic at infinity*.

Let us now discuss each of the three categories.

a) We know that a *doubly ruled* projective quadric either intersects a given plane, or it is tangent to it. This is also true for $\underline{\mathbf{H}}$. In the case of an intersection, the resulting affine quadric will be a hyperboloid,

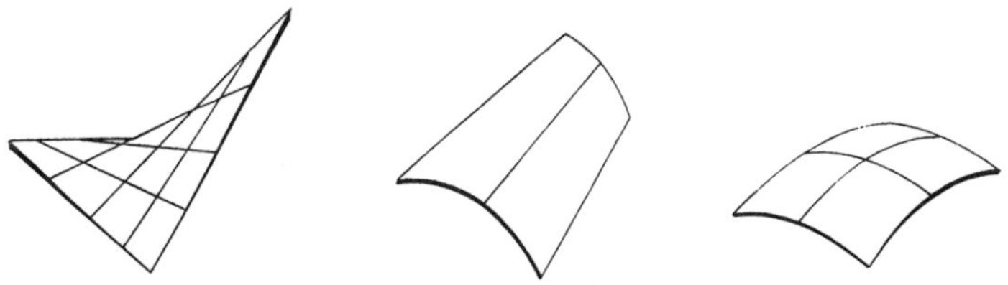

Figure 13.2. Quadric categories: a) doubly ruled, b) cone, c) oval.

called *hyperboloid of one sheet.* In the case of tangency, we obtain a paraboloid, namely the *hyperbolic paraboloid.*

b) A singly ruled quadric may be of three types, depending on the type of the conic at infinity. We distinguish between hyperbolic, parabolic, and elliptic cones. In the special case that all straight lines on the cone are parallel, we have hyperbolic, parabolic, and elliptic *cylinders.* The parabolic cylinder really is a paraboloid, touching the plane at infinity along a straight line.

c) A projective *oval* quadric may intersect $\underline{\mathbf{H}}$ in a conic; then the corresponding affine quadric is a hyperboloid and is called *hyperboloid of two sheets.* If the projective oval quadric touches $\underline{\mathbf{H}}$, the resulting affine conic is a paraboloid; it is called *elliptic paraboloid.* If the projective oval quadric has no point in common with $\underline{\mathbf{H}}$, the resulting affine quadric is simply called an *ellipsoid.* It is the only finite affine quadric.

One can list seventeen types of affine quadrics — but for that, one has to count all kinds of degeneracies. For example, a pair of parallel planes and a pair of intersecting planes constitute two more examples of quadrics. For details, see the literature on projective and/or analytic geometry, for example [16], [25], [66], [111].

13.3 The Stereographic Projection

We are interested in the relationship between Bézier patches and quadrics; in order to address that problem, we need a tool: the *stereographic projection.* Let $\underline{\mathbf{Q}}$ denote a quadric surface, and let $\underline{\mathbf{c}}$ be a point on it, and further let $\underline{\mathbf{P}}$ be an arbitrary plane not through $\underline{\mathbf{c}}$. See Figure 13.3.

If $\underline{\mathbf{x}}$ is an arbitrary point on the quadric, we may form the line through $\underline{\mathbf{c}}$ and $\underline{\mathbf{x}}$ and intersect it with the plane $\underline{\mathbf{P}}$, resulting in a point $\hat{\underline{\mathbf{x}}}$. We say that $\hat{\underline{\mathbf{x}}}$ is the image of $\underline{\mathbf{x}}$ under the stereographic projection with center $\underline{\mathbf{c}}$.

There are two important geometric properties of stereographic maps:

a) They provide a one-to-one map between the quadric and the plane, except for one singular point.

b) The image of a conic on the quadric is a conic on the plane.

The most important *algebraic* property of stereographic maps is that they can be described by quadratic maps. To make this notion more precise, we follow a development by Teller and Séquin [113]. Referring to Figure 13.4, let $\underline{\mathbf{b}}_{002}, \underline{\mathbf{b}}_{020}$, and $\underline{\mathbf{b}}_{200}$ be three points on the quadric. We choose

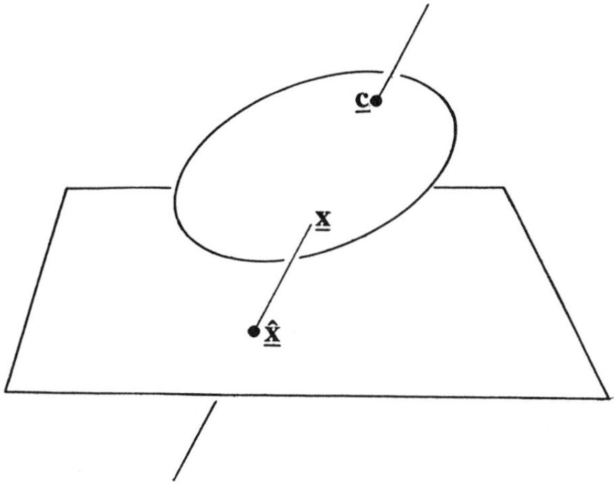

Figure 13.3. Stereographic projections: a point on a quadric is projected onto a plane, with the center of the projection being on the quadric.

this naming convention as we shall make these three points the corners of a triangular quadratic Bézier patch. The stereographic projection of these three points yields a triangle in the plane, which will serve as the domain for the triangular patch.

Our desired Bézier patch will have three planar boundary curves, and for these we choose the planes through $\underline{\mathbf{c}}$ and the three corner points $\underline{\mathbf{b}}_{002}, \underline{\mathbf{b}}_{020}$, and $\underline{\mathbf{b}}_{200}$. Each of these planes cuts a conic out of the quadric, and these conics will be our patch boundary curves. Let us concentrate on the boundary curve $v = 0$, lying in the plane spanned by $\underline{\mathbf{c}}, \underline{\mathbf{b}}_{002}, \underline{\mathbf{b}}_{200}$. It will be of the form

$$\underline{\mathbf{b}}(t) = \underline{\mathbf{b}}_{002}B_0^2(t) + \underline{\mathbf{b}}_{101}B_1^2(t) + \underline{\mathbf{b}}_{200}B_2^2(t). \tag{13.3}$$

We can find the Bézier point $\mathbf{b}_{101}$ by intersecting the plane of the boundary curve with the tangent planes at $\underline{\mathbf{b}}_{002}$ and $\underline{\mathbf{b}}_{200}$; see Figure 13.5. We pick the shoulder tangent such that $\underline{\mathbf{c}}$ corresponds to $t = \infty$.

The remaining boundary curves should have the same form as (13.3). This is achieved by also selecting $\underline{\mathbf{c}}$ as corresponding to the parameter value ∞ for all boundary curves.

We now have completed the construction of a quadratic patch, interpolating to data from the quadric. In fact, this patch *is* the quadric since it agrees with the quadric in three points and the tangent planes there, plus in another point $\underline{\mathbf{c}}$.

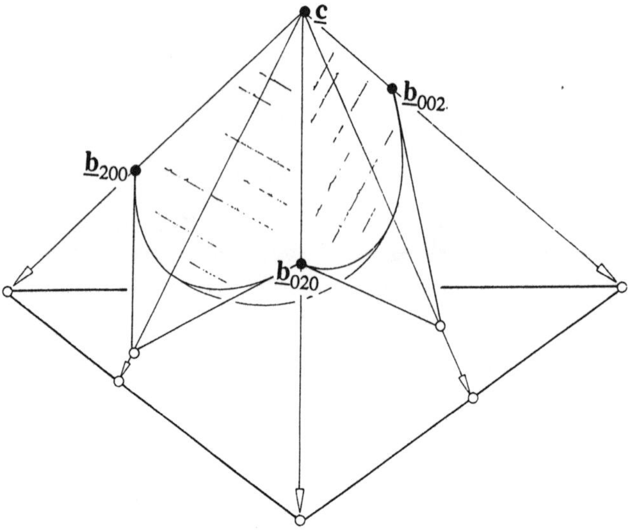

Figure 13.4. Stereographic projections: a quadratic Bézier triangle can be utilized to describe the map.

In our development, we have not only constructed a patch on the quadric, but we have also fixed a coordinate system on the plane: let $\underline{\mathbf{a}}_{ijk}$ be the projection of $\underline{\mathbf{b}}_{ijk}$ through $\underline{\mathbf{c}}$ onto the plane. Then the three lines $\underline{\mathbf{a}}_{002} \wedge \underline{\mathbf{a}}_{110}$, $\underline{\mathbf{a}}_{020} \wedge \underline{\mathbf{a}}_{101}$, and $\underline{\mathbf{a}}_{200} \wedge \underline{\mathbf{a}}_{011}$ are concurrent, their intersection $\underline{\mathbf{a}}_{111}$ being the projection of the point $\underline{\mathbf{b}}(\frac{1}{3}, \frac{1}{3}, \frac{1}{3})$. The four points $\underline{\mathbf{a}}_{002}, \underline{\mathbf{a}}_{020}, \underline{\mathbf{a}}_{200}$, and $\underline{\mathbf{a}}_{111}$ thus form a projective reference frame for the plane: every point $\underline{\mathbf{a}}$ in the plane can be written as $\underline{\mathbf{a}} = u\underline{\mathbf{a}}_{200} + v\underline{\mathbf{a}}_{020} + w\underline{\mathbf{a}}_{002}$. The point $\underline{\mathbf{b}}(u, v, w)$ is then the image of $\underline{\mathbf{a}}$.

To summarize: starting with a quadric, a point on it, and a plane, we defined a stereographic projection. We have then shown that this stereographic projection can be interpreted as a quadratic map of the plane to the quadric, i.e., every point $\underline{\mathbf{x}}$ on the quadric can be written as

$$\underline{\mathbf{x}} = \underline{\mathbf{x}}(u, v, w) = \sum_{|\mathbf{i}|=2} \mathbf{b}_{\mathbf{i}} B^2_{\mathbf{i}}(u, v, w).$$

Note that this quadratic map is not unique: it depends on the choice of the three patch corners.

We conclude this section with some further (projective) analysis of a quadratic patch on a quadric. The projective reference frame in the plane defines a fundamental line $\underline{\mathbf{F}}$ in the plane, as illustrated earlier in Figure

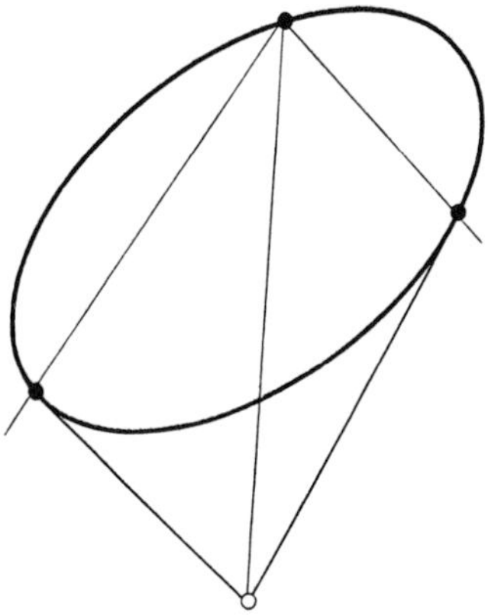

Figure 13.5. Stereographic projections: a boundary curve is determined by its Bézier points and a shoulder tangent.

1.9. It contains the points $(-\infty, \infty, 0), (-\infty, 0, \infty)$, and $(0, -\infty, \infty)$. Now recall that the point $\underline{\mathbf{c}}$ on the quadric corresponded to just those values, for that was how we parametrized the patch boundary curves. It follows that the whole fundamental line is mapped to $\underline{\mathbf{c}}$ by the stereographic projection! In other words, the stereographic projection is not one-to-one for $\underline{\mathbf{F}}$.

It then follows that the plane defined by $\underline{\mathbf{c}}$ and $\underline{\mathbf{F}}$ is the quadric's tangent plane at $\underline{\mathbf{c}}$; see Figure 13.6 for an illustration of the case of a cone.

Quadratic surfaces have been studied in detail by J. Steiner; thus one also encounters the name "Steiner patches" for quadratic Bézier triangles; see Sederberg [105].

13.4 Quartics on Quadrics

Given a quadric in $\mathbb{P}^3$, a point on it, and a plane, we could find quadratic patches on the quadric by means of stereographic projections. If we do not refer to the stereographic projection, we may describe the resulting quadratic patches as follows:

a) The three boundary curves meet in one point $\underline{\mathbf{c}}$.

b) The three boundary curves are coplanar at $\underline{\mathbf{c}}$.

c) Each boundary curve assumes the parameter value $\pm\infty$ at $\underline{\mathbf{c}}$.

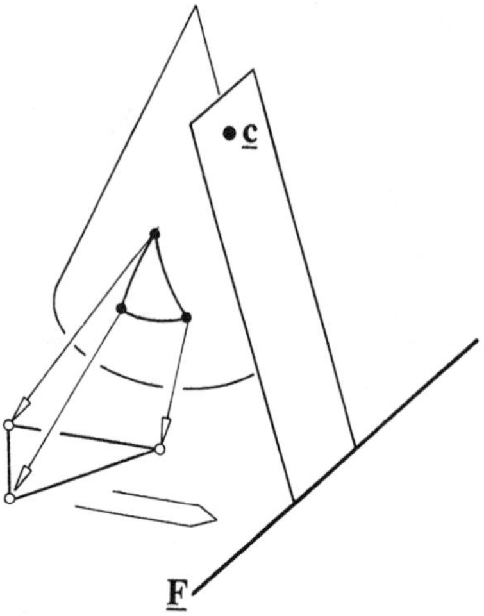

Figure 13.6. Stereographic projections: the tangent plane at the center intersects the projection plane in the fundamental line.

So if we encounter a quadratic patch that satisfies the above three conditions, we may conclude that it is a patch on a quadric: the information provided is sufficient to uniquely determine a quadric. A warning is in place: the above conditions cannot be checked in a numerically reliable way for many cases!

Not all quadratic patches are on quadrics, since not all quadratic patches meet the three conditions. In terms of algebraic geometry, a quadratic patch is said to have *algebraic degree* four, meaning that a quadratic patch and a straight line may have up to four intersections.[2] A quadric, on the other hand, can only have two such intersections.

If we are given three planar sections of a quadric (remember: planar sections of quadrics are always conics!) as shown in Figure 13.7, can we find a triangular patch with those conics as boundaries that is on the given quadric? The answer is yes, but the patch will not be of degree two — unless, of course, the above three conditions are met.

The stereographic projection may again be used to find the desired patch formulation. Pick any point $\underline{c}$ on the quadric, and an arbitrary plane not through $\underline{c}$. Project the three boundary curves, together with their control polygons, into the plane, using a projection with center $\underline{c}$. The images of the conics will be three conics, as shown in Figure 13.7. Let the plane have

[2]Counting coincident as well as complex intersections.

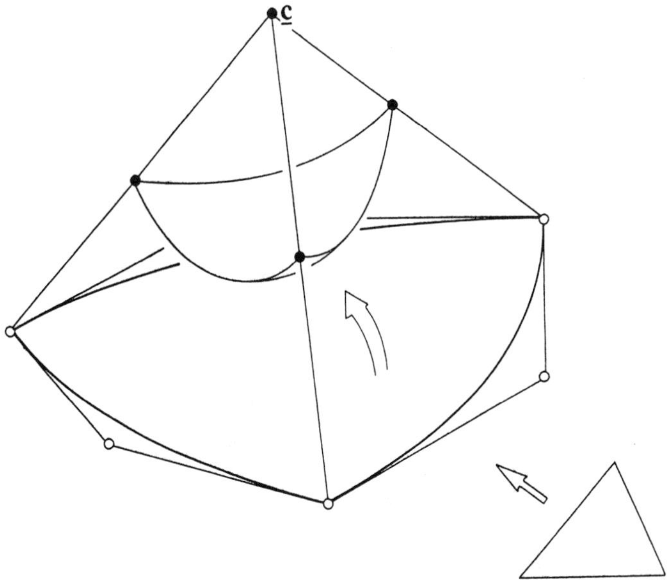

Figure 13.7. Nonquadratic patches on quadrics: the three boundary curves may be projected onto three conics in a plane.

projective coordinates u, v, w.

The three conics may be interpreted as the boundary curves of a planar quadratic patch $\underline{\mathbf{a}}$. It is the image of a triangle with projective coordinates (r, s, t):

$$\underline{\mathbf{a}}(r, s, t) = \sum_{|\mathbf{i}|=2} \underline{\mathbf{a}}_{\mathbf{i}} B_{\mathbf{i}}^2(r, s, t).$$

This quadratic Bézier triangle is a map of the (r, s, t) plane onto the (u, v, w) plane. We know that the stereographic projection is a quadratic map (once we picked three points on the quadric and proceeded as in Section 13.3). Thus our triangular patch on the quadric is the composition of the two quadratic maps, making it a degree four patch, a quartic.[3]

There is one important difference between this kind of quartic patch versus the quadratic one discussed above: the quadratic patch covers the quadric completely, and, with the exception of $\underline{\mathbf{c}}$, provides a one-to-one coverage. The quartic, on the other hand, does neither: it does not cover

[3] To avoid a terminology overload, the basic definitions at a glance: quadratic: being of degree two; quartic: being of degree four; quadric: a surface with algebraic degree two.

the whole quadric; some parts of the quadric are covered twice, others not at all. The reason is that the patch $\underline{\mathbf{u}} = \underline{\mathbf{a}}(r, s, t)$ does not cover all of $\mathbb{P}^2$; it may be interpreted as a "flattened out" quadric itself. This phenomenon was observed by M. Niebuhr [84].

Any triangular patch on a quadric can be written as a degree four Bézier triangle, provided it has planar boundary curves. Once projected into affine space, the quartic patch becomes a rational quartic, and so all triangular patches (with planar boundary curves) on affine quadrics can be written as rational quartic Bézier triangles. In the special case of a sphere, Section 15.6 gives an explicit example.

In the same manner, one may obtain a rectangular biquartic patch on the sphere, see Cobb [31]. Its Bézier points and their weights are listed in Section 15.7. Six such patches (suitably rotated) cover the sphere with a rectangular network that has a cube topology.

For more literature regarding rational patches on quadrics, see also [70], [69].

13.5 Problems

1. If a rational cubic triangular patch (with its domain extended to be the whole extended affine plane) is *closed*, does this mean it is a quadric?

2. Find a rational quadratic patch that has four intersections with some straight line.

3. A triangle divides the projective plane into four regions, see Section 1.10. Mark these regions on a quadric that is defined by a quadratic Bézier triangle.

4. Given a rational quadratic patch on a quadric (i.e., one that satisfies the three conditions from Section 13.4), devise a method to decide its type – doubly ruled, singly ruled, or oval.

14

Gregory Patches

Bicubic patches (integral ones) have enjoyed enormous popularity. Another type of surface, the Coons patch, never became quite as popular because of problems people had with a correct specification of certain twist vectors (see [47] for details). The twist problem was first fixed by J. Gregory in 1972, still in the context of Coons patches. What really made his idea recognized and successful was, however, a modification by Chiyokura and Kimura as described in [29] and [28]. The following sections will focus on that approach; for the original literature, see Barnhill [13], Barnhill and Gregory [14], and Gregory [65] [1].

J. Gregory has contributed more to the field of Geometric Modeling than just the patches that now bear his name. He also obtained fundamental results in the areas of interpolatory subdivision, n-sided patches, and shape preserving interpolation. John Gregory died in March 1993, at the age of 47.

[1] In the earlier literature, Gregory patches are typically referred to as "Gregory squares".

14.1 Bicubic Gregory Patches

Consider the problem that is illustrated by Figure 14.1: two (integral) bicubic Bézier patches A and B are given. A third one is sought such that we achieve C^1 continuity across all patch boundaries.

The C^1 conditions between rectangular Bézier patches determine all points in the desired patch — with one exception: the pair of points indicated by solid arrows. Of those two, the point $\mathbf{b}_{1,1}^{(u)}$ is determined by patch A, while $\mathbf{b}_{1,1}^{(v)}$ is determined by patch B. Obviously, no bicubic Bézier patch can have two different points $\mathbf{b}_{1,1}$, and so the given problem has no solution. The way out of this dilemma is the inception of a rational patch that *does* incorporate two different points $\mathbf{b}_{1,1}$, namely both $\mathbf{b}_{1,1}^{(u)}$ and $\mathbf{b}_{1,1}^{(v)}$.

The key idea is to make $\mathbf{b}_{1,1}$ a function of u and v:

$$\mathbf{b}_{11} = \mathbf{b}_{11}(u,v) = \frac{u\mathbf{b}_{11}^{(v)} + v\mathbf{b}_{11}^{(u)}}{u+v}.$$

We also define

$$\begin{aligned}
\mathbf{b}_{21} &= \frac{(1-u)\mathbf{b}_{21}^{(v)} + v\mathbf{b}_{21}^{(u)}}{(1-u)+v},\\
\mathbf{b}_{12} &= \frac{u\mathbf{b}_{12}^{(v)} + (1-v)\mathbf{b}_{12}^{(u)}}{u+(1-v)},\\
\mathbf{b}_{22} &= \frac{(1-u)\mathbf{b}_{22}^{(v)} + (1-v)\mathbf{b}_{22}^{(u)}}{(1-u)+(1-v)}.
\end{aligned}$$

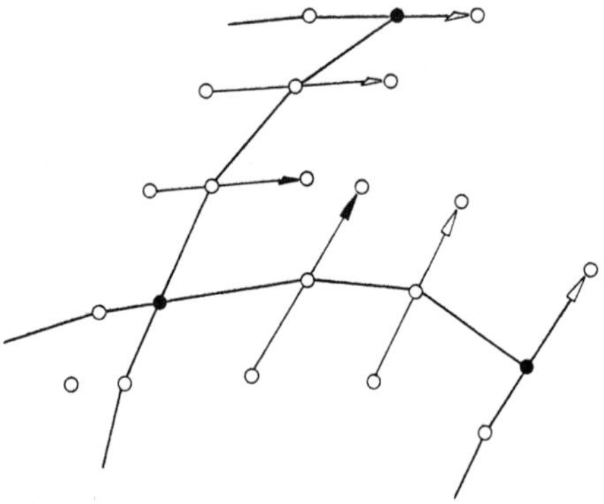

Figure 14.1. Gregory patches: the given data (solid circles) are incompatible.

This is the Chiyokura/Kimura approach, based on the original approach by J. Gregory. It replaces the terms $\mathbf{b}_{11}$, etc. in the original Bézier formulation by the above terms. Note that if $\mathbf{b}_{11}^{(u)} = \mathbf{b}_{11}^{(v)}$, etc., the bicubic Gregory patch becomes a standard bicubic Bézier patch.

What have we gained? Let us take v-partials of the Gregory patch along the boundary curve $v = 0$. We obtain

$$\begin{aligned} \mathbf{x}_v(u,0) &= 3 * \big[(1-u)^3[\mathbf{b}_{01} - \mathbf{b}_{00}] + 3(1-u)^2 u[\mathbf{b}_{11}^{(u)} - \mathbf{b}_{10}] \\ &\quad +3(1-u)u^2[\mathbf{b}_{21}^{(u)} - \mathbf{b}_{20}] + u^3[\mathbf{b}_{31} - \mathbf{b}_{30}]\big]. \end{aligned} \tag{14.1}$$

In words: the cross boundary derivative $\mathbf{x}_v(u,0)$ only depends on the control points as shown in Figure 14.2. Thus it solves the incompatibility problem highlighted in Figure 14.1, since analogous statements can be made for the remaining boundary curves.

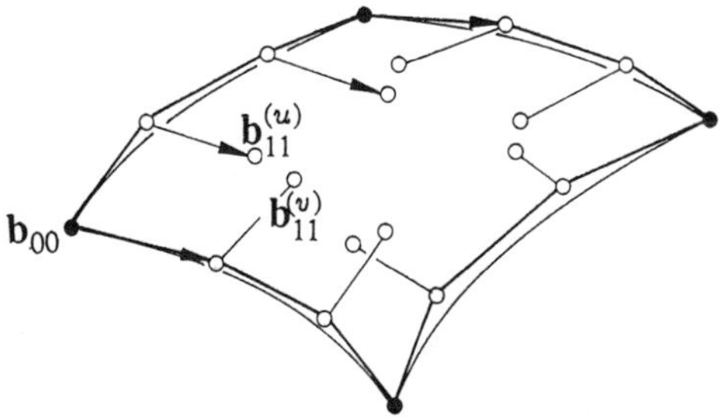

Figure 14.2. Gregory patches: the cross boundary derivative along $u = 0$ only depends on the indicated difference vectors.

It is now time to describe another important property of Gregory patches: their corner twists[2] are *discontinuous*. To make this claim more precise, consider the corner twist $\mathbf{x}_{uv}(0,0)$. When we try to compute it, we end up with an expression of the form $\mathbf{0}/0$, a meaningless expression. But we may evaluate $\mathbf{x}_{uv}(0,0)$ by taking certain limits. If we approach $(0,0)$ along $u = 0$, we obtain

$$\lim_{u \to 0} \mathbf{x}_{uv}(0,0) = 9[\mathbf{b}_{11}^{u} - \mathbf{b}_{10} - \mathbf{b}_{01} + \mathbf{b}_{00}].$$

Taking a different limit, by approaching $(0,0)$ along $v = 0$, we obtain

$$\lim_{v \to 0} \mathbf{x}_{vu}(0,0) = 9[\mathbf{b}_{11}^{v} - \mathbf{b}_{10} - \mathbf{b}_{01} + \mathbf{b}_{00}].$$

[2]These are the patches' mixed partials, evaluated at the corners.

14.2 The Rational Bézier Form

The appearance of terms such as $u/(u+v)$ in our Gregory patches make them *rational.* However, they are not of the form (11.3). But after some rewriting, one obtains a rational Bézier patch of degree seven in both u and v. Quoting [112], the Bézier points $\mathbf{c}_{ij}$:

$$
\begin{aligned}
\mathbf{c}_{00} &= \mathbf{b}_{00},\\
\mathbf{c}_{01} &= \mathbf{b}_{00},\\
\mathbf{c}_{02} &= (5\mathbf{b}_{00} + 6\mathbf{b}_{01})/11,\\
\mathbf{c}_{03} &= (2\mathbf{b}_{00} + 15\mathbf{b}_{01} + 6\mathbf{b}_{02})/23,\\
\mathbf{c}_{11} &= (7\mathbf{b}_{00} + 3\mathbf{b}_{01} + 3\mathbf{b}_{01})/13,\\
\mathbf{c}_{12} &= (18\mathbf{b}_{11}^{(v)} + 28\mathbf{b}_{00} + 42\mathbf{b}_{01} + 6\mathbf{b}_{02} + 15\mathbf{b}_{10})/109,\\
\mathbf{c}_{13} &= (45\mathbf{b}_{11}^{(v)} + 18\mathbf{b}_{12}^{(u)} + 14\mathbf{b}_{00} + 84\mathbf{b}_{01} + 42\mathbf{b}_{02} + 2\mathbf{b}_{03} + 6\mathbf{b}_{10})/211\\
\mathbf{c}_{22} &= (63\mathbf{b}_{11}^{(u)} + 63\mathbf{b}_{11}^{(v)} + 18\mathbf{b}_{12}^{(v)} + 18\mathbf{b}_{21}^{(u)} + 52\mathbf{b}_{00} + 84\mathbf{b}_{01} + 15\mathbf{b}_{02} + 84\mathbf{b}_{10}\\
&\quad + 15\mathbf{b}_{20})/412\\
\mathbf{c}_{23} &= (63\mathbf{b}_{11}^{(u)} + 189\mathbf{b}_{11}^{(v)} + 18\mathbf{b}_{12}^{(u)} + 108\mathbf{b}_{12}^{(v)}\\
&\quad + 27\mathbf{b}_{21}^{(u)} + 18\mathbf{b}_{21}^{(v)} + 9\mathbf{b}_{22}^{(u)} + 9\mathbf{b}_{22}^{(v)} + 28\mathbf{b}_{00}\\
&\quad + 156\mathbf{b}_{01} + 84\mathbf{b}_{02} + 5\mathbf{b}_{03} + 42\mathbf{b}_{10} + 6\mathbf{b}_{13} + 6\mathbf{b}_{20})/703\\
\mathbf{c}_{33} &= (234\mathbf{b}_{11}^{(u)} + 234\mathbf{b}_{11}^{(v)} + 108\mathbf{b}_{12}^{(u)} + 144\mathbf{b}_{12}^{(v)} + 144\mathbf{b}_{21}^{(u)} + 108\mathbf{b}_{21}^{(u)}\\
&\quad + 63\mathbf{b}_{22}^{(u)} + 14\mathbf{b}_{00} + 84\mathbf{b}_{01} + 42\mathbf{b}_{02} + 2\mathbf{b}_{03} + 84\mathbf{b}_{10} + 15\mathbf{b}_{13} + 42\mathbf{b}_{20}\\
&\quad + 6\mathbf{b}_{23} + 2\mathbf{b}_{30} + 15\mathbf{b}_{31}6\mathbf{b}_{32})/1064
\end{aligned}
$$

and corresponding weights w_{ij}:

$$
\begin{aligned}
w_{00} &= 0,\\
w_{01} &= 2/7,\\
w_{02} &= 11/21,\\
w_{03} &= 23/35,\\
w_{11} &= 26/49,\\
w_{12} &= 109/147\\
w_{13} &= 211/245\\
w_{22} &= 412/441\\
w_{23} &= 256/245\\
w_{33} &= 282/245.
\end{aligned}
$$

The remaining control points and weights follow by symmetry. The conversion to Bézier form is advantageous for algorithms that benefit from the convex hull property of rational Bézier patches or that need a subdivision process. Note, however, that these patches cannot be formulated in the IGES specification, since that only allows for strictly positive weights. A way around this is described by Ueda [114].

Gregory patches degenerate to $\mathbf{0}/0$ at the patch corners — this is mirrored by the appearance of zero weights for the corner Bézier points. These singularities are removable: the patch evaluates to the original corner Bézier points.

Parameter values that evaluate to $\mathbf{0}/0$ are called *base points.* These singularities are not always removable: in our case, we have $\mathbf{c}_{01} = \mathbf{c}_{10}$, which is also the patch's corner position. If $\mathbf{c}_{01} \neq \mathbf{c}_{10}$, $(u,v) = (0,0)$ would be mapped to all values on the straight line between $\mathbf{c}_{01}$ and $\mathbf{c}_{10}$ — a nonremovable singularity. Which value on the line is attained depends on the direction from which $(0,0)$ is approached as a limit. More on base points in Section 14.5.

14.3 RBG Patches

Gregory patches replace the "twist" control points of a bicubic patch by variable points, thus introducing rational terms. One can go a step further and repeat this approach for rational Bézier patches. Then *every* control point and its corresponding weight would be made variable. Such an approach seems overly liberal for the boundary curve control points — it seems reasonable to have those determined in a non variable fashion. Still, each control point might have two weights. This was carried out by H. Chiyokura et al. in [30]. The corresponding method is called *rational boundary Gregory patch.*

For an explanation, let us group the 16 control points of a rational bicubic into four blocks, one per patch corner. For the block pertaining to $\mathbf{b}_{00}$, we have the following:

$$\mathbf{b}_{i,j} = \frac{u^2\mathbf{b}_{i,j}^{(u)} + v^2\mathbf{b}_{i,j}^{(v)}}{u^2+v^2}; \quad i,j \in \{0,1\} \tag{14.2}$$

and

$$w_{i,j} = \frac{u^2 w_{i,j}^{(u)} + v^2 w_{i,j}^{(v)}}{u^2+v^2}; \quad i,j \in \{0,1\} \tag{14.3}$$

and analogous requirements for the remaining three blocks. Equations (14.2) actually contain only one "real" equation, namely for $(i,j) = (1,1)$, since we required all boundary control points to be constant.

An additional requirement is that the weights $w_{i,j}^{(u)}$ are equal for the first and last two rows of control points:

$$w_{i,0}^{(u)} = w_{i,1}^{(u)} \text{ and } w_{i,2}^{(u)} = w_{i,3}^{(u)}$$

for $i = 0, 1, 2, 3$. Similarly, it is required that

$$w_{0,j}^{(v)} = w_{1,j}^{(u)} \text{ and } w_{2,j}^{(v)} = w_{3,j}^{(u)}.$$

These conditions allow for easy formulation of C^1 conditions between neighboring patches: the cross boundary derivative along the edge $v = 0$ is given by

$$\mathbf{x}_v(u,0) = \frac{3\sum_{i=0}^{3} B_i^3(u) w_{i,0}^{(u)} [\mathbf{b}_{i,1}^{(u)} - \mathbf{b}_{i,0}^{(u)}]}{\sum_{i=0}^{3} B_i^3(u) w_{i,0}^{(u)}}. \tag{14.4}$$

The RBG concept has been successfully implemented into the DESIGN-BASE modeler; see Chiyokura [28].

14.4 Gregory Triangles

Before he invented the rectangular patch that now bears his name, J. Gregory investigated triangular patches.[3] We will present a Bézier based version of that patch scheme, as described in [80] and [56].

Consider three triangular cubic patches as in Figure 14.3. If patches A and C are given, can we determine the central control point $\mathbf{b}_{111}$ of patch B so as to ensure C^1 continuity of the overall surface? This is not possible; $\mathbf{b}_{111}$ would be overdetermined.

The Foley/Opitz solution: replace $\mathbf{b}_{111}$ by three independent points $\mathbf{b}_{111}^{(u)}, \mathbf{b}_{111}^{(v)}, \mathbf{b}_{111}^{(w)}$! In the equation for the patch, we will then replace the constant point $\mathbf{b}_{111}$ by the function

$$\mathbf{b}_{111}(u,v,w) = \frac{vw\mathbf{b}_{111}^{(u)} + uw\mathbf{b}_{111}^{(v)} + uv\mathbf{b}_{111}^{(w)}}{vw + uw + uv}. \tag{14.5}$$

Again, cross boundary derivatives across the edge $u = 0$ will only involve one of the three center points, namely $\mathbf{b}_{111}^{(u)}$, with similar statements for the remaining two edges.

It is possible to convert a Gregory triangle into standard rational Bézier form — it will then be a rational triangular patch of degree five. Its Bézier

[3]Private communication. It is also interesting to note that de Casteljau considered triangular patches before he turned to rectangular ones!

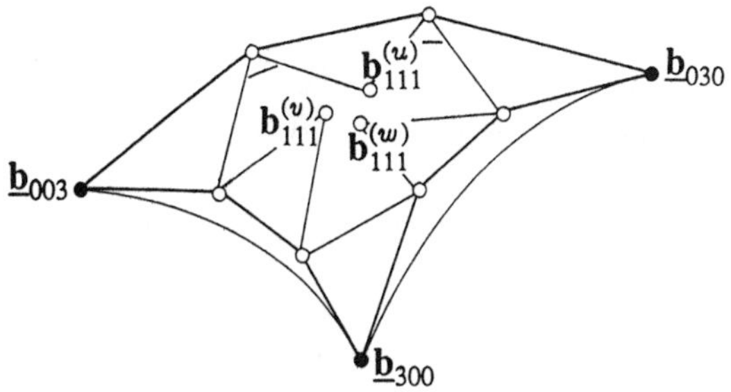

Figure 14.3. Gregory triangles: standard cubic patches have compatibility problems.

points $\mathbf{c}_{ijk}$ are

$$\begin{aligned}
\mathbf{c}_{050} &= \mathbf{b}_{030},\\
\mathbf{c}_{041} &= \mathbf{b}_{030},\\
\mathbf{c}_{032} &= \mathbf{b}_{021},\\
\mathbf{c}_{131} &= (\mathbf{b}_{030} + 3\mathbf{b}_{021} + 3\mathbf{b}_{120})/7,\\
\mathbf{c}_{122} &= (\mathbf{b}_{021} + \mathbf{b}_{012} + 2\mathbf{b}^{(u)}_{111})/4.
\end{aligned}$$

The corresponding weights w_{ijk} are:

$$\begin{aligned}
w_{050} &= 0,\\
w_{041} &= 1/5,\\
w_{032} &= 3/10,\\
w_{131} &= 7/20,\\
w_{122} &= 2/5.
\end{aligned}$$

All other control points and weights follow by symmetry.

Again, we note base points at the patch corners! The patch has a removable singularity there since the control points $\mathbf{c}_{041}$ and $\mathbf{c}_{140}$ coincide, as do the corresponding ones at the other corners.

The approach by Longhi [80] uses quartic triangles and replaces each of the three interior control points by a weighted combination of two points, thus yielding a total of 18 control points. We have that

$$\mathbf{b}_{211}(u,v,w) = \frac{(1-v)w\mathbf{b}^{(w)}_{211} + v(1-w)\mathbf{b}^{(v)}_{211}}{(1-v)w + v(1-w)}$$

and analogous expressions for $\mathbf{b}_{121}$ and $\mathbf{b}_{112}$.

14.5 Base Points and Multisided Patches

When transcribing a Gregory patch into rational Bézier form, we generated a triangular patch of the following structure (we only describe the corner $\mathbf{c}_{0n0}$, the other ones being similar): the corner weight w_{0n0} is zero, and the next two control points $\mathbf{c}_{1,n-1,0}$ and $\mathbf{c}_{0,n-1,1}$ coincide. If they did not, then our patch would contain the straight line $\overline{\mathbf{c}_{1,n-1,0}, \mathbf{c}_{0,n-1,1}}$ and similar straight lines at the other corners. Then we would have a six-sided patch! Three of its boundary "curves" would just be straight lines, of course, while the remaining three would be of degree $n-2$.

We may add more layers of zero weights: if we also set $w_{1,n-1,0} = w_{0,n-1,1} = 0$, then the next layer of control points; $\mathbf{c}_{2,n-2,0}, \mathbf{c}_{1,n-2,1}, \mathbf{c}_{0,n-1,2}$ and their weights, would generate a rational quadratic boundary curve of the six-sided patch. This approach of creating multisided patches using base points is due to J. Warren [116]. Figure 14.4 illustrates. These multisided patches can be very useful for purposes of filleting and blending.

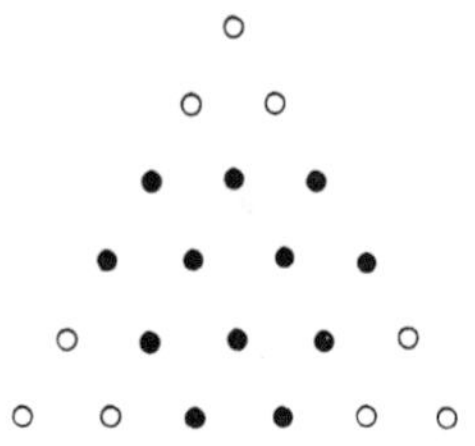

Figure 14.4. Multisided patches: by setting some weights zero (open circles), the remaining control points (solid circles) form a six-sided patch. Three of its boundary curves are of degree two, and three are of degree one.

14.6 Problems

1. Plot the function $z = u/(u+v)$. Study how its behavior near $(0,0)$ explains the properties of Gregory patches.

2. Then repeat the same for the function $z = vw/(vw + uw + uv)$, but now with u, v, w being barycentric coordinates of a triangle in the plane $z = 0$.

3. Suppose the boundary lines of a bilinear patch do not meet in the patch corners, but miss each other. Devise a bilinear Gregory patch, i.e., one that interpolates to eight "corner points."

15

Examples and Standards

This chapter contains practical information for NURBS implementers. It contains the most important data specifications, namely two IGES formats, and several examples of commonly encountered NURBS curves and surfaces.

15.1 IGES

One of the big obstacles in the CAD/CAM world used to be the diversity of data formats that were used by different systems. Ford used Coons patches, General Motors used Gordon surfaces, and Mercedes used bicubic splines, to name just a few. In recent years, there has been a trend towards unification — many systems now produce geometry that adheres to one common data standard: IGES, short for initial graphics exchange specification. IGES defines most conceivable geometric objects, from points, lines, and planes to transformation matrices; here we are only interested in their NURBS formats. More advanced data exchange formats are presently under way, the most important one being STEP, which aims beyond pure geometry: a product is not simply defined by its shape, but also by variables such as material, or surface finish. STEP integrates such descriptors. Further down the road we should expect specifications that involve variable dimensions, such as the radii of cylinders in a complex assembly.

name	type	description
N	int	number of control points (counted from 0)
n	int	degree
flag1	int	0: nonplanar, 1: planar
flag2	int	0: open, 1: closed
flag3	int	0: rational, 1: integral
flag4	int	0: nonperiodic, 1: periodic
		knot sequence
u_{-n}	double	first knot
$\vdots$	$\vdots$	$\vdots$
u_{N+1}	double	last knot
		weights
w_0	double	first weight
$\vdots$	$\vdots$	$\vdots$
w_N	double	last weight
		control points
$\mathbf{d}_0^x$	double	x-component of first control point
$\mathbf{d}_0^y$	double	y-component of first control point
$\mathbf{d}_0^z$	double	z-component of first control point
$\vdots$	$\vdots$	$\vdots$
$\mathbf{d}_N^x$	double	x-component of last control point
$\mathbf{d}_N^y$	double	y-component of last control point
$\mathbf{d}_N^z$	double	z-component of last control point
		range
u_{start}	double	starting parameter value
u_{end}	double	ending parameter value
		plane normal (if flag1=1)
$\mathbf{n}^x$	double	x-component of normal
$\mathbf{n}^y$	double	y-component of normal
$\mathbf{n}^z$	double	z-component of normal

IGES format 126: NURBS curves

The IGES NURBS curve format #126 invites several comments:

- All of the knots may occur with multiplicity greater than one. If an interior knot is listed k times, the curve is only guaranteed to be $n-k$ times differentiable there. For example, a G^2 cubic spline curve from Chapter 9 will require double knots; but after a reparametrization to C^2, simple knots will suffice.

- If the end knots u_n and u_{N+1} are of multiplicity n, the first and last control points will lie on the curve.
- IGES requires $n + 1$ "dummy" knots at the beginning and end of the knot sequence. The first and last of these have no influence on any computation – they are an IGES design flaw.
- For periodic curves, IGES offers too many knots, as the beginning and end of the curve is no more special than any point in the interior.
- IGES only allows strictly positive weights; this will cause problems with control vectors (Chapter 5) as well as with periodic curves with simple knots.
- IGES stores 3-D points and associated weights; this is *not* the homogeneous form! In order to obtain it, we would have to multiply every point by its corresponding weight.
- IGES also provides a "Directory Entry" in which special properties may be stored, such as being part of a circle, a hyperbola, etc.

name	type	description
M	int	number of control points (counted from 0)in u-direction
N	int	number of control points (counted from 0)in v-direction
m	int	degree in u
n	int	degree in v
flag1	int	1: closed in u-direction, 0: else
flag2	int	1: closed in v-direction, 0: else
flag3	int	0: rational, 1: integral
flag4	int	0: nonperiodic in u, 1: periodic
flag5	int	0: nonperiodic in v, 1: periodic
		knot sequences
u_{-m}	double	first knot in u
$\vdots$	$\vdots$	$\vdots$
u_{M+1}	double	last knot in u
v_{-n}	double	first knot in v
$\vdots$	$\vdots$	$\vdots$
v_{N+1}	double	last knot in v
		weights
$w_{0,0}$	double	first weight
$w_{1,0}$	double	second weight
$\vdots$	$\vdots$	$\vdots$
$w_{M,N}$	double	last weight
		control points
$\mathbf{d}^x_{0,0}$	double	x-component of first control point
$\mathbf{d}^y_{0,0}$	double	y-component of first control point
$\mathbf{d}^z_{0,0}$	double	z-component of first control point
$\mathbf{d}^x_{1,0}$	double	x-component of second control point
$\mathbf{d}^y_{1,0}$	double	y-component of second control point
$\mathbf{d}^z_{1,0}$	double	z-component of second control point
$\vdots$	$\vdots$	$\vdots$
$\mathbf{d}^x_{M,N}$	double	x-component of last control point
$\mathbf{d}^y_{M,N}$	double	y-component of last control point
$\mathbf{d}^z_{M,N}$	double	z-component of last control point
		ranges
u_{start}	double	starting parameter value in u
u_{end}	double	ending parameter value in u
v_{start}	double	starting parameter value in v
v_{end}	double	ending parameter value in v

IGES format 128: NURBS surfaces

Some comments on the IGES 128 surface format:

- Since only strictly positive weights are allowed, control vectors (Chapter 5) or Gregory patches (Chapter 14) will cause problems.
- There is a "Directory Entry" to record special features such as being a right cylinder, a torus, or a surface of revolution.
- For issues concerning knot multiplicities, see the comments for the curve format.
- Triangular patches (Chapter 12) are not supported.

15.2 A Semicircle

A semicircle may be written as a C^2 piecewise cubic NURB curve with two segments. Assuming a uniform knot sequence, its homogeneous control points are given by:

$$\begin{bmatrix} -1 \\ 0 \\ 1 \end{bmatrix}, \frac{2}{3}\begin{bmatrix} -1 \\ \frac{1}{2} \\ 1 \end{bmatrix}, \frac{1}{3}\begin{bmatrix} 0 \\ 2 \\ 1 \end{bmatrix}, \frac{2}{3}\begin{bmatrix} 1 \\ \frac{1}{2} \\ 1 \end{bmatrix}, \begin{bmatrix} 1 \\ 0 \\ 1 \end{bmatrix}.$$

In the IGES format, we would have to enter the knots as: 0,0,0,1,2,2,2.

15.3 A Hyperbolic Paraboloid

A triangular patch on a hyperbolic paraboloid is given by

$$\begin{matrix} & & \begin{bmatrix} 0 \\ 1 \\ 0 \\ 1 \end{bmatrix} & & \\ & \begin{bmatrix} 0 \\ \frac{1}{2} \\ 0 \\ 1 \end{bmatrix} & & \begin{bmatrix} \frac{1}{2} \\ \frac{1}{2} \\ \frac{1}{2} \\ 1 \end{bmatrix} & \\ \begin{bmatrix} 0 \\ 0 \\ 0 \\ 1 \end{bmatrix} & & \begin{bmatrix} \frac{1}{2} \\ 0 \\ 0 \\ 1 \end{bmatrix} & & \begin{bmatrix} 1 \\ 0 \\ 0 \\ 1 \end{bmatrix} \end{matrix}.$$

This patch is purely polynomial. Of course, it may also be represented as a *bilinear* patch with control points

$$\begin{bmatrix} 0 \\ 1 \\ 0 \\ 1 \end{bmatrix} \begin{bmatrix} 1 \\ 1 \\ 1 \\ 1 \end{bmatrix} \begin{bmatrix} 0 \\ 0 \\ 0 \\ 1 \end{bmatrix} \begin{bmatrix} 1 \\ 0 \\ 0 \\ 1 \end{bmatrix}.$$

Note that the two diagonals $\mathbf{b}(u, u)$ and $\mathbf{b}(u, 1-u)$ are parabolas, touching the plane at infinity.

15.4 A Quarter of a Cylinder

A quarter of a cylinder, with the z−axis as its axis, is given by the homogeneous biquadratic control net

$$\begin{bmatrix} 1 \\ 0 \\ 0 \\ 1 \end{bmatrix} \sqrt{2}/2 \begin{bmatrix} 1 \\ 1 \\ 0 \\ 1 \end{bmatrix} \begin{bmatrix} 1 \\ 0 \\ 0 \\ 1 \end{bmatrix}$$

$$\begin{bmatrix} 1 \\ 0 \\ 1 \\ 1 \end{bmatrix} \sqrt{2}/2 \begin{bmatrix} 1 \\ 1 \\ 1 \\ 1 \end{bmatrix} \begin{bmatrix} 1 \\ 0 \\ 1 \\ 1 \end{bmatrix}.$$

If we let all points with z-component equal to 1 collapse into one point $[0, 0, 1, 1]^{\mathrm{T}}$, we obtain a quarter of a cone.

15.5 A Rational Quadratic Patch on a Sphere

The following six homogeneous control points and weights (quoted from Niebuhr [84]) describe a rational quadratic triangular patch on a sphere:

$$\begin{bmatrix} 0 \\ 0 \\ 1 \\ 1 \end{bmatrix}$$

$$\begin{bmatrix} \frac{1}{2}\sqrt{2} \\ 0 \\ \frac{1}{2}\sqrt{2} \\ \frac{1}{2}\sqrt{2} \end{bmatrix} \begin{bmatrix} \frac{1}{2} \\ \frac{1}{2} \\ \frac{1}{2} \\ \frac{1}{2} \end{bmatrix}$$

$$\begin{bmatrix} 1 \\ 0 \\ 0 \\ 1 \end{bmatrix} \begin{bmatrix} \frac{1}{2}\sqrt{2} \\ \frac{1}{2}\sqrt{2} \\ 0 \\ \frac{1}{2}\sqrt{2} \end{bmatrix} \begin{bmatrix} 0 \\ 1 \\ 0 \\ 1 \end{bmatrix}.$$

Such patches are obtainable from a stereographic projection; in this case, its center $\underline{\mathbf{c}}$ is given by

$$\underline{\mathbf{c}} = \begin{bmatrix} -1 \\ 0 \\ 0 \\ 1 \end{bmatrix}.$$

15.6 An Octant of a Sphere

We give a rational quartic representation of an octant of the sphere as a triangular patch (Farin, Piper, Worsey [49]). The control net (in homogeneous form) is given by

$$w_{040} \begin{bmatrix} 0 \\ 0 \\ 1 \\ 1 \end{bmatrix}$$

$$w_{031} \begin{bmatrix} \alpha \\ 0 \\ 1 \\ 1 \end{bmatrix} \quad w_{130} \begin{bmatrix} 0 \\ \alpha \\ 1 \\ 1 \end{bmatrix}$$

$$w_{022} \begin{bmatrix} \beta \\ 0 \\ \beta \\ 1 \end{bmatrix} \quad w_{121} \begin{bmatrix} \gamma \\ \gamma \\ 1 \\ 1 \end{bmatrix} \quad w_{220} \begin{bmatrix} 0 \\ \beta \\ \beta \\ 1 \end{bmatrix}$$

$$w_{013}\begin{bmatrix}1\\0\\\alpha\\1\end{bmatrix} w_{112}\begin{bmatrix}1\\\gamma\\\gamma\\1\end{bmatrix} w_{211}\begin{bmatrix}\gamma\\1\\\gamma\\1\end{bmatrix} w_{310}\begin{bmatrix}0\\1\\\alpha\\1\end{bmatrix}$$

$$w_{004}\begin{bmatrix}1\\0\\0\\1\end{bmatrix} w_{103}\begin{bmatrix}1\\\alpha\\0\\1\end{bmatrix} w_{202}\begin{bmatrix}\beta\\\beta\\0\\1\end{bmatrix} w_{301}\begin{bmatrix}\alpha\\1\\0\\1\end{bmatrix} w_{400}\begin{bmatrix}0\\1\\0\\1\end{bmatrix}$$

where

$$\alpha = (\sqrt{3}-1)/\sqrt{3},\quad \beta = (\sqrt{3}+1)/2\sqrt{3},\quad \gamma = 1-(5-\sqrt{2})(7-\sqrt{3})/46.$$

The weights are:

$$w_{040} = 4\sqrt{3}(\sqrt{3}-1), w_{031} = 3\sqrt{2}, w_{202} = 4, w_{121} = \frac{\sqrt{2}}{\sqrt{3}}(3+2\sqrt{2}-\sqrt{3}),$$

the other ones following by symmetry.

In order to represent the whole sphere, we would assemble eight copies of this octant patch. Other representations are also possible: each octant may be written as a rational biquadratic patch (introducing singularities at the north and south poles); see [92].

15.7 A Sixth of a Sphere

The following is due to J. Cobb [31]. The sphere is covered by six rectangular rational biquartic patches, creating a cube-like topology. Each of those patches may be obtained by appropriately rotating the given control net. We give the homogeneous form.

$$\begin{bmatrix}4(1-t)\\4(1-t)\\4(1-t)\\4(3-t)\end{bmatrix}\begin{bmatrix}-d\\d(t-4)\\d(t-4)\\d(3t-2)\end{bmatrix}\begin{bmatrix}0\\4(1-2t)/3\\4(1-2t)/3\\4(5-t)/3\end{bmatrix}\begin{bmatrix}d\\d(t-4)\\d(t-4)\\d(3t-2)\end{bmatrix}\begin{bmatrix}4(t-1)\\4(1-t)\\4(1-t)\\4(3-t)\end{bmatrix}$$

$$\begin{bmatrix}d(t-4)\\-d\\d(t-4)\\d(3t-2)\end{bmatrix}\begin{bmatrix}(2-3t)/2\\(2-3t)/2\\-(t+6)/2\\(t+6)/2\end{bmatrix}\begin{bmatrix}0\\d(2t-7)/3\\-5\sqrt{6}/3\\d((t+6)/3\end{bmatrix}\begin{bmatrix}(3t-2)/2\\(2-3t)/2\\-(t+6)/2\\(t+6)/2\end{bmatrix}\begin{bmatrix}d(4-t)\\-d\\d(t-4)\\d(3t-2)\end{bmatrix}$$

$$\begin{bmatrix}4(1-2t)/3\\0\\4(1-2t)/3\\4(5-t)/3\end{bmatrix}\begin{bmatrix}d(2t-7)/3\\0\\-5\sqrt{6}/3\\d(t+6)/3\end{bmatrix}\begin{bmatrix}0\\0\\4(t-5)/3\\4(5t-1)/9\end{bmatrix}\begin{bmatrix}-d(2t-7)/3\\0\\-5\sqrt{6}/3\\d(t+6)/3\end{bmatrix}\begin{bmatrix}-4(1-2t)/3\\0\\4(1-2t)/3\\4(5-t)/3\end{bmatrix}$$

$$\begin{bmatrix} d(t-4) \\ -d \\ d(t-4) \\ d(3t-2) \end{bmatrix} \begin{bmatrix} (2-3t)/2 \\ -(2-3t)/2 \\ -(t+6)/2 \\ (t+6)/2 \end{bmatrix} \begin{bmatrix} 0 \\ -d(2t-7)/3 \\ -5\sqrt{6}/3 \\ d((t+6)/3 \end{bmatrix} \begin{bmatrix} (3t-2)/2 \\ -(2-3t)/2 \\ -(t+6)/2 \\ (t+6)/2 \end{bmatrix} \begin{bmatrix} d(4-t) \\ d \\ d(t-4) \\ d(3t-2) \end{bmatrix}$$

$$\begin{bmatrix} 4(1-t) \\ -4(1-t) \\ 4(1-t) \\ 4(3-t) \end{bmatrix} \begin{bmatrix} -d \\ -d(t-4) \\ d(t-4) \\ d(3t-2) \end{bmatrix} \begin{bmatrix} 0 \\ -4(1-2t)/3 \\ 4(1-2t)/3 \\ 4(5-t)/3 \end{bmatrix} \begin{bmatrix} d \\ -d(t-4) \\ d(t-4) \\ d(3t-2) \end{bmatrix} \begin{bmatrix} 4(t-1) \\ -4(1-t) \\ 4(1-t) \\ 4(3-t) \end{bmatrix}$$

where $d = \sqrt{2}$ and $t = \sqrt{3}$.

15.8 A Sixteenth of a Torus

A torus is a surface of revolution. We consider a torus that has been created by revolving a circle around the z-axis. Half of the torus is above the $z = 0$ plane, the other half is below it. Here are the homogeneous control points for an "interior" patch with $x, y, z \geq 0$, writing it in homogeneous biquadratic form:

$$\begin{bmatrix} 1 \\ 0 \\ 0 \\ 1 \end{bmatrix} \begin{bmatrix} \frac{1}{2}\sqrt{2} \\ \frac{1}{2}\sqrt{2} \\ 0 \\ \frac{1}{2}\sqrt{2} \end{bmatrix} \begin{bmatrix} 0 \\ 1 \\ 0 \\ 1 \end{bmatrix} \begin{bmatrix} \frac{1}{2}\sqrt{2} \\ 0 \\ \frac{1}{2}\sqrt{2} \\ \frac{1}{2}\sqrt{2} \end{bmatrix} \begin{bmatrix} \frac{1}{2} \\ \frac{1}{2} \\ \frac{1}{2} \\ \frac{1}{2} \end{bmatrix} \begin{bmatrix} 0 \\ \frac{1}{2}\sqrt{2} \\ \frac{1}{2}\sqrt{2} \\ \frac{1}{2}\sqrt{2} \end{bmatrix} \begin{bmatrix} 2 \\ 0 \\ 0 \\ 1 \end{bmatrix} \begin{bmatrix} \sqrt{2} \\ \sqrt{2} \\ \frac{1}{2}\sqrt{2} \\ \frac{1}{2}\sqrt{2} \end{bmatrix} \begin{bmatrix} 0 \\ 2 \\ 1 \\ 1 \end{bmatrix}$$

Just for the record: a torus may have four intersections with a straight line – hence it is *not* a quadric!

15.9 Problems

1. The hyperbolic paraboloid from Section 15.3 is obtainable from a stereographic projection. What is its center?
2. Redo Section 15.3 and the previous problem for an *elliptic paraboloid.*
3. Write a semisphere as a rational triangular cubic patch.

Notation

Here is the notation used in this book:

$\wedge$	cross product; $\underline{\mathbf{L}} \wedge \underline{\mathbf{M}}$: intersection of two lines; $\underline{\mathbf{p}} \wedge \underline{\mathbf{q}}$: line through two points
$\dot{}\,, \ddot{}$	curve derivatives with respect to the current parameter
a, b, α, β	real numbers or real-valued functions
$\mathbf{0}$	short for (0,0,0)
$\mathbf{a}, \mathbf{b}$	points or vectors in affine space
$\underline{\mathbf{a}}, \underline{\mathbf{b}}$	points in projective space If $\underline{\mathbf{b}} = [x, y, z, w]^{\mathrm{T}}$, then $\mathbf{b} = [x/w, y/w, z/w]^{\mathrm{T}}$.
A, B	matrices
$\underline{\mathbf{L}}, \underline{\mathbf{M}}$	lines in projective space
B_i^n, B_{i}^n	Bernstein polynomials of degree n
$\mathbf{e1}, \mathbf{e2}, \mathbf{e3}$	short for $[1, 0, 0]^{\mathrm{T}}, [0, 1, 0]^{\mathrm{T}}, and [0, 0, 1]^{\mathrm{T}}$
$\mathbb{E}^d$	d-dimensional Euclidean space
$D_{\mathbf{d}} f$	directional derivative of f in the direction $\mathbf{d}$
Δ_i	difference in parameter intervals (i.e., $\Delta_i = u_{i+1} - u_i$)
Δ^r	iterated forward difference
H_i^3	cubic Hermite polynomials
$\mathbf{P}$	control polygon
Φ	an affine or projective map
$\|\|\mathbf{v}\|\|$	(Euclidean) length of the vector $\mathbf{v}$
$\mathbf{x}_u$	u-partial of $\mathbf{x}(u, v)$

Bibliography

[1] S. Abhyankar and C. Bajaj. Automatic parametrization of rational curves and surfaces I: conics and conicoids. *Computer Aided Design*, 19(1), 1987.

[2] S. Abhyankar and C. Bajaj. Automatic parametrization of rational curves and surfaces II: cubics and cubicoids. *Computer Aided Design*, 19(9), 1987.

[3] S. Abhyankar and C. Bajaj. Automatic parametrization of rational curves and surfaces III: Algebraic plane curves. *Computer Aided Geometric Design*, 5(4):309–322, 1988.

[4] S. Abhyankar and C. Bajaj. Automatic parametrization of rational curves and surfaces IV. *ACM Transactions on Graphics*, 8(4):325–334, 1989.

[5] D. Ahuja and S. Coons. Geometry for construction and display. *IBM System Journal*, 7(3/4):188–205, 1968.

[6] G. Albrecht. A remark on Farin points for Bézier triangles. *Computer Aided Geometric Design*, 11, 1994.

[7] G. Aumann. Interpolation with developable Bézier patches. *Computer Aided Geometric Design*, 8(5):409–420, 1991.

[8] H. Baker. *Principles of Geometry, Vol. III.* Cambridge Univ. Press, London, 1934.

[9] A. Ball. Consurf I: introduction of the conic lofting tile. *Computer Aided Design*, 6(4):243–249, 1974.

[10] A. Ball. Consurf II: description of the algorithms. *Computer Aided Design*, 7(4):237–242, 1975.

[11] A. Ball. Consurf III: how the program is used. *Computer Aided Design*, 9(1):9–12, 1977.

[12] L. Bardis and N. Patrikalakis. Approximate conversion of rational B-spline patches. *Computer Aided Geometric Design*, 6(3):189–204, 1989.

[13] R. Barnhill. Coons' patches. *Computers in Industry*, 3:37–43, 1982.

[14] R. Barnhill and J. Gregory. Compatible smooth interpolation in triangles. *J of Approx. Theory*, 15(3):214–225, 1975.

[15] R. Bartels, J. Beatty, and B. Barsky. *An Introduction to Splines for Use in Computer Graphics and Geometric Modeling.* Morgan Kaufmann, San Mateo, 1987.

[16] M. Berger. *Geometry I.* Springer-Verlag, Berlin-Heidelberg-New York, 1987.

[17] D. Binschadler and W. Frey. Developable surfaces for math-based binder design, 1993. GMR report MA-526.

[18] J. Blinn and M. Newell. Clipping using homogeneous coordinates. *Computer Graphics*, 12(3):281–287, 1978.

[19] R. Bodduluri and B. Ravani. Design of developable surfaces using duality between plane and point geometries. *Computer Aided Design*, 25(10):621–632, 1993.

[20] W. Boehm. Inserting new knots into B-spline curves. *Computer Aided Design*, 12(4):199–201, 1980.

[21] W. Boehm. On cubics: a survey. *Computer Graphics and Image Processing*, 19:201–226, 1982.

[22] W. Boehm. Rational geometric splines. *Computer Aided Geometric Design*, 4(1-2):67–77, 1987.

[23] W. Boehm. An affine representation of de Casteljau's and de Boor's rational algorithms. *Computer Aided Geometric Design*, 10(10):175–180, 1993.

[24] W. Boehm and D. Hansford. Bézier patches on quadrics. In G. Farin, editor, *NURBS for Curve and Surface Design*, pages 1–14. SIAM, Philadelphia, 1991.

[25] W. Boehm and H. Prautzsch. *Geometric Concepts for Geometric Design.* AK Peters, Ltd., Wellesley, 1994.

[26] G. Bol. *Projective Differential Geometry, Vol. 1.* Vandenhoeck and Ruprecht, Goettingen, 1950. Vol. 2 in 1954, Vol.3 in 1967. In German.

[27] K. Bolton. Biarc curves. *Computer Aided Design*, 7(2):89–92, 1975.

[28] H. Chiyokura. *Solid Modeling with DESIGNBASE.* Addison-Wesley, 1988.

[29] H. Chiyokura and F. Kimura. Design of solids with free-form surfaces. *Computer Graphics*, 17(3):289–298, 1983.

[30] H. Chiyokura, T. Takamura, K. Konno, and T. Harada. G^1 surface interpolation over irregular meshes with rational curves. In G. Farin, editor, *NURBS for Curve and Surface Design*, pages 15–34. SIAM, 1991.

[31] J. Cobb. A rational bicubic representation of the sphere. Technical report, Computer science, Univ. of Utah, 1988.

[32] E. Cohen, T. Lyche, and R. Riesenfeld. Discrete B-splines and subdivision techniques in computer aided geometric design and computer graphics. *Comp. Graphics and Image Process.*, 14(2):87–111, 1980.

[33] H. Coxeter. *The Real Projective Plane.* Cambridge Univ. Press, Cambridge, 1961.

[34] P. Davis. *Interpolation and Approximation.* Dover, New York, 1975. First edition 1963.

[35] C. de Boor. On calculating with B-splines. *J Approx. Theory*, 6(1):50–62, 1972.

[36] C. de Boor. *A Practical Guide to Splines.* Springer, New York-Berlin-Heidelberg, 1978.

[37] C. de Boor, K. Hollig, and M. Sabin. High accuracy geometric Hermite interpolation. *Computer Aided Geometric Design*, 4(4):269–278, 1987.

[38] P. de Casteljau. Outillages méthodes calcul. Technical report, A. Citroen, Paris, 1959.

[39] P. de Casteljau. Courbes et surfaces à poles. Technical report, A. Citroen, Paris, 1963.

[40] M. do Carmo. *Differential Geometry of Curves and Surfaces.* Prentice Hall, Englewood Cliffs, 1976.

[41] M. Eck. Degree reduction of Bézier curves. *Computer Aided Geometric Design*, 10(3-4):237–252, 1993.

[42] G. Farin. Konstruktion und Eigenschaften von Bézier-Kurven und -Flächen. Master's thesis, Technical University Braunschweig, FRG, 1977.

[43] G. Farin. Algorithms for rational Bézier curves. *Computer Aided Design*, 15(2):73–77, 1983.

[44] G. Farin. Triangular Bernstein-Bézier patches. *Computer Aided Geometric Design*, 3(2):83–128, 1986.

[45] G. Farin. Curvature continuity and offsets for piecewise conics. *ACM Transactions on Graphics*, 8(2):89–99, 1989.

[46] G. Farin. Rational curves and surfaces. In T. Lyche and L. Schumaker, editors, *Mathematical Aspects in CAGD*, pages 215–238. Academic Press, Boston, 1989.

[47] G. Farin. *Curves and Surfaces for Computer Aided Geometric Design.* Academic Press, Boston, 1992. Third edition.

[48] G. Farin. Tighter convex hulls for rational Bézier curves. *Computer Aided Geometric Design*, 10(2):123–126, 1993.

[49] G. Farin, B. Piper, and A. Worsey. The octant of a sphere as a non-degenerate triangular Bézier patch. *Computer Aided Geometric Design*, 4(4):329–332, 1988.

[50] G. Farin and H. Pottmann. Projective algorithms for developable surfaces, 1994. Submitted for publication.

[51] R. Farouki and T. Sakkalis. Real rational curves are not 'unit speed'. *Computer Aided Geometric Design*, 8(2):151–158, 1991.

[52] J. Fiorot and P. Jeannin. *Rational Curves and Surfaces.* Wiley, Chicester, 1992. Translated from the French by M. Harrison.

[53] M. Floater. Evaluation and properties of the derivatives of a NURBS curve. In T. Lyche and L. Schumaker, editors, *Mathematical Methods in CAGD II*, pages 261–274. Academic Press, Boston, 1992.

[54] M. Floater. Derivatives of rational Bézier curves. *Computer Aided Geometric Design*, 10, 1993.

[55] J. Foley and A. Van Dam. *Fundamentals of Interactive Computer Graphics.* Addison-Wesley, Reading, 1982.

[56] T. Foley and K. Opitz. Hybrid cubic Bézier patches. In T. Lyche and L. Schumaker, editors, *Mathematical Methods in CAGD II*, pages 275–286. Academic Press, Boston, 1992.

[57] A. Forrest. *Curves and surfaces for computer-aided design.* PhD thesis, Cambridge Univ., 1968.

[58] A. Forrest. Interactive interpolation and approximation by Bézier polynomials. *The Computer J*, 15(1):71–79, 1972. Reprinted in *Computer Aided Design*, 22(9):527-537,1990.

[59] A. Forrest. The twisted cubic curve: a computer-aided geometric design approach. *Computer Aided Design*, 12(4):165–172, 1980.

[60] G. Geise and B. Juettler. A geometrical approach to curvature continuous joints of rational curves. *Computer Aided Geometric Design*, 10(2):109–122.

[61] G. Geise and B. Juettler. The constructive geometry of Bézier curves, 1991. Manuscript.

[62] R. Goldman. The method of resolvents: a technique for the implicitization, inversion, and intersection of non-planar, parametric, rational cubic curves. *Computer Aided Geometric Design*, 2(4):237–255, 1985.

[63] R. Goldman, T. Sederberg, and D. Anderson. Vector elimination: A technique for the implicitization, inversion, and intersection of planar parametric rational polynomial curves. *Computer Aided Geometric Design*, 1(4):327–356, 1984.

[64] T. Goodman. Shape preserving interpolation by parametric rational cubic splines. Technical report, University of Dundee, 1988. Department of Mathematics and Computer Science.

[65] J. Gregory. Smooth interpolation without twist constraints. In R. E. Barnhill and R. F. Riesenfeld, editors, *Computer Aided Geometric Design*, pages 71–88. Academic Press, Boston, 1974.

[66] D. Hilbert and S. Cohn-Vossen. *Geometry and the Imagination.* Chelsea, New York, 1952.

[67] M. Hohmeyer and B. Barsky. Rational continuity: parametric and geometric continuity of rational polynomial curves. *ACM Transactions on Graphics*, 8(4):335–359, 1989.

[68] J. Hoschek. Dual Bézier curves and surfaces. In R. Barnhill and W. Boehm, editors, *Surfaces in Computer Aided Geometric Design*, pages 147–156. North-Holland, 1983.

[69] J. Hoschek. Bézier curves and surfaces on quadrics. pages 1–4. Academic Press, Boston, 1992.

[70] J. Hoschek and D. Lasser. *Grundlagen der Geometrischen Datenverarbeitung.* B.G. Teubner, Stuttgart, 1989. English translation: *Fundamentals of Computer Aided Geometric Design*, A K Peters,Ltd., Wellesley, 1993.

[71] J. Hoschek and F. Schneider. Spline conversion for trimmed rational Bézier- and B-spline surfaces. *Computer Aided Design*, 22(9):580–590, 1990.

[72] J. Hoschek and F.-J. Schneider. Approximate spline conversion for integral and rational Bézier and B-Spline surfaces. In R. E. Barnhill, editor, *Geometry Processing for Design and Manufacturing*, pages 45–86. SIAM, Philadelphia, 1992.

[73] S. Jolles. *Die Theorie der Oskulanten und das Sehnensystem der Raumkurve 4. Ordnung, 2. Spezies.* PhD thesis, Technical Univ. Aachen, 1886.

[74] R. Klass. An offset spline approximation for plane cubics. *Computer Aided Design*, 15(5):296–299, 1983.

[75] M. Lachance. Chebychev economization for parametric surfaces. *Computer Aided Geometric Design*, 5(3):195–208, 1988.

[76] M. Lachance and A. Schwartz. Four point parabolic interpolation. *Computer Aided Geometric Design*, 8(2):143–150, 1991.

[77] E. Lee and M. Lucian. Moebius reparametrizations of rational B-splines. *Computer Aided Geometric Design*, 8(3):213–216, 1991.

[78] R. Liming. *Practical analytical geometry with applications to aircraft.* Macmillan, New York, 1944.

[79] R. Liming. *Mathematics for Computer Graphics.* Aero Publishers, 1979.

[80] L. Longhi. Interpolating patches between cubic boundaries. Technical report, Univ. of California, Berkeley, 1986. Technical Report T.R. UCB/CSD 87/313.

[81] M. Lucian. Linear fractional transformations of rational Bézier curves. In G. Farin, editor, *NURBS for Curve and Surface Design*, pages 131–139. SIAM, Philadelphia, 1991.

[82] T. Lyche. Note on the Oslo algorithm. *Computer Aided Design*, 20(6):353–355, 1988.

[83] F. Moebius. *August Ferdinand Moebius, Gesammelte Werke.* Verlag von S. Hirzel, 1885. Also published by Dr. M. Saendig oHG, Wiesbaden, FRG, 1967.

[84] M. Niebuhr. *Properties of quadrics in rational Bézier form.* PhD thesis, TU Braunschweig, Germany, 1992. In German.

[85] D. Parkinson and D. Moreton. Optimal biarc-curve fitting. *Computer Aided Design*, 23(6):411–419, 1991.

[86] N. Patrikalakis. Approximate conversion of rational splines. *Computer Aided Geometric Design*, 6(2):155–165, 1989.

[87] D. Pedoe. *A course of Geometry.* Cambridge Univ. Press, London, 1970.

[88] M. Penna and R. Patterson. *Projective Geometry and its Applications to Computer Graphics.* Prentice Hall, Englewood Cliffs, 1986.

[89] L. Piegl. On the use of infinite control points in CAGD. *Computer Aided Geometric Design*, 4(1-2):155–166, 1987.

[90] L. Piegl. Modifying the shape of rational B-splines. Part 1: curves. *Computer Aided Design*, 21(8):509–518, 1989.

[91] L. Piegl. On NURBS: a survey. *Computer Graphics and Applications*, 11(1):55–71, 1990.

[92] L. Piegl and W. Tiller. Curve and surface constructions using rational B-splines. *Computer Aided Design*, 19(9):485–498, 1987.

[93] H. Pottmann. Locally controllable conic splines with curvature continuity. *ACM Transactions on Graphics*, 10(4):366–377, 1991.

[94] H. Pottmann. A projectively invariant characterization of G^2 continuity for rational curves. In G. Farin, editor, *NURBS for Curve and Surface Design*, pages 141–148. SIAM, Philadelphia, 1991.

[95] H. Prautzsch. A short proof of the Oslo algorithm. *Computer Aided Geometric Design*, 1(1):95–96, 1984.

[96] L. Ramshaw. Blossoming: a connect-the-dots approach to splines. Technical report, Digital Systems Research Center, Palo Alto, 1987.

[97] L. Ramshaw. Blossoms are polar forms. *Computer Aided Geometric Design*, 6(4):323–359, 1989.

[98] P. Redont. Representation and deformation of developable surfaces. *Computer Aided Design*, 21(1):13–20, 1989.

[99] R. Riesenfeld. Homogeneous coordinates and projective planes in computer graphics. *IEEE Computer Graphics and Applications*, 1:50–55, 1981.

[100] M. Rowin. Conic, cubic, and T-conic segments. Technical Report Document D2-23252, The Boeing Co., 1964.

[101] M. Sabin. *The use of piecewise forms for the numerical representation of shape.* PhD thesis, Hungarian Academy of Sciences, Budapest, Hungary, 1976.

[102] L. Schumaker. *Spline functions: Basic Theory.* Wiley, New York, 1981.

[103] T. Sederberg. *Implicit and parametric curves and surfaces for computer aided geometric design.* PhD thesis, Mech. Eng., Purdue Univ., 1983.

[104] T. Sederberg. Improperly parametrized rational curves. *Computer Aided Geometric Design*, 3(1):67–75, 1986.

[105] T. Sederberg and D. Anderson. Steiner surface patches. *IEEE Computer Graphics and Applications*, 5(5):23–36, 1985.

[106] T. Sederberg, D. Anderson, and R. Goldman. Implicit representation of parametric curves and surfaces. *Computer Vision, Graphics, and Image Processing*, 28(1):72–84, 1984.

[107] T. Sederberg and M. Kakimoto. Approximating rational curves using polymial curves. In G. Farin, editor, *NURBS for Curve and Surface Design*, pages 149–158. SIAM, 1991.

[108] T. Sederberg and X. Wang. Rational hodographs. *Computer Aided Geometric Design*, 4(4):333–335, 1987.

[109] T. Sharrock. Biarcs in three dimensions. In R. Martin, editor, *The Mathematics of Surfaces II*, pages 395–412. Oxford University Press, Oxford, 1987.

[110] J. Stolfi. *Oriented Projective Geometry.* Academic Press, Boston, 1991.

[111] D. Struik. *Analytic and Projective Geometry.* Addison-Wesley, Cambridge, MA, 1953.

[112] T. Takamura, M. Ohta, H. Toriya, and H. Chiyokura. A method to convert a Gregory patch and rational boundary Gregory patch to a rational Bézier patch and its applications. In T. Chua and T. Kunij, editors, *Proceedings of Computer Graphics International '90*, pages 543–562. Springer, Berlin-Heidelberg-New York, 1990.

[113] S. Teller and C. Sequin. Modeling implicit quadrics and free-form surfaces with trimmed rational quadratic Bézier patches. Technical report, Computer Science, Berkeley, 1990. Report no. UCB/CSD 90/577.

[114] K. Ueda. A method for removing the singularities from Gregory surfaces. In T. Lyche and L. Schumaker, editors, *Mathematical Methods in CAGD II.* Academic Press, Boston, 1992.

[115] K. Vesprille. *Computer aided design applications of the rational B-spline approximation form.* PhD thesis, Syracuse Univ., 1975.

[116] J. Warren. Creating multisided rational Bézier surfaces using base points. *ACM Transactions on Graphics*, 11(2):127–139, 1992.

[117] M. Watkins and A. Worsey. Degree reduction for Bézier curves. *Computer Aided Design*, 20(7):398–405, 1988.

[118] J. Wilczynski. *Projective Differential Geometry of Curves and Ruled Surfaces.* Chelsea, New York, 1961.

[119] A. Worsey, T. Sederberg, and G. Wang. A recursive algortitm for Hermite approximation of rational curves, 1994. Submitted for publication.

Index